D0946687

MICROTECHNOLOGY AND MEMS

Springer
Berlin
Heidelberg
New York
Hong Kong
London
Milan
Paris
Tokyo

Physics and Astronomy

http://www.springer.de/phys/

MICROTECHNOLOGY AND MEMS

Series Editor: H. Baltes   D. Liepmann

The series Microtechnology and MEMS comprises text books, monographs, and state-of-the-art reports in the very active field of microsystems and microtechnology. Written by leading physicists and engineers, the books describe the basic science, device design, and applications. They will appeal to researchers, engineers, and advanced studends.

Mechanical Microsensors
By M. Elwenspoek and R. Wiegerink

CMOS Cantilever Sensor Systems
Atomic Force Microscopy and Gas Sensing Applications
By D. Lange, O. Brand, and H. Baltes

Micromachines as Tools for Nanotechnology
Editor: H. Fujita

Modelling of Microfabrication Systems
By R. Nassar and W. Dai

F.W. Olin College Library

Raja Nassar, Weizhong Dai

Modelling of Microfabrication Systems

With 140 Figures

 Springer

F.W. Olin College Library

Prof. Raja Nassar
Prof. Weizhong Dai

Louisiana Technical University
Faculty of Mathematics and Statistics
71272 Ruston, LA
USA

e-mail: Nassar@coes.latech.edu
       Dai@coes.latech.edu

ISBN 3-540-00252-9 Springer-Verlag Berlin Heidelberg New York

Cataloging-in-Publication Data applied for.
Bibliographic information published by Die Deutsche Bibliothek. Die Deutsche Bibliothek lists this publication in the Deutsche Nationalbibliografie; detailed bibliographic data is available in the Internet at <http://dnb.ddb.de>.

This work is subject to copyright. All rights are reserved, whether the whole or part of the material is concerned, specifically the rights of translation, reprinting, reuse of illustrations, recitation, broadcasting, reproduction on microfilm or in other ways, and storage in data banks. Duplication of this publication or parts thereof is permitted only under the provisions of the German Copyright Law of September 9, 1965, in its current version, and permission for use must always be obtained from Springer-Verlag. Violations are liable to prosecution under German Copyright Law.

Springer-Verlag Berlin Heidelberg New York
a member of BertelsmannSpringer Science+Business Media GmbH

http://www.springer.de

© Springer-Verlag Berlin Heidelberg 2003
Printed in Germany

The use of general descriptive names, registered names, trademarks, etc. in this publication does not imply, even in the absence of a specific statement, that such names are exempt from the relevant protective laws and regulations and therefore free for general use.

Typesetting: Datenconversion by author
Cover-design: design & production, Heidelberg
Printed on acid-free paper   62 / 3020 hu – 5 4 3 2 1 0

This book is dedicated to Ita, Mark, Chuhan, and Sijie.

Preface

Microtechnology is a rapidly growing area for research and development. The importance of this technology is becoming increasingly apparent and its potential is enormous in terms of new products and product enhancements. Bulk micromachining and surface micromachining are the main fabrication techniques that have been used to manufacture a variety of microsensors (such as pressure and acceleration sensors) that blossomed into a vital industry with numerous applications in the aerospace/military/avionic industry, the medical industry, and the automotive industry, as well as in certain consumer products. Techniques used in the fabrication of microsensors have some drawbacks in the sense that the structures produced are planar and somewhat fragile. For further development of the technology and its industrial market, it is necessary to develop new technology to enable the manufacturing of high aspect ratio (height/width) microparts. This technology is needed to improve the rigidity of devices and allow coupling to them, thus enabling the manufacture of complex mechanisms such as microrelays, actuators, and micromotors. Equipment with high aspect ratio would revolutionize many types of microelectromechanical systems (MEMS), including sensors now on the market (Bryzek and Peterson 1994; Wise and Najafi 1991).

There are several important prototyping techniques or processes that are being developed at present for the manufacture of high aspect ratio microstructures. These include ion beam micromachining, x-ray lithography, laser chemical vapor deposition, laser photopolymerization, and laser ablation. Mathematical modeling of these microfabrication systems has played an important role in their development. A model that can accurately predict the behavior of a system is essential for its control and optimization. The purpose of this book is to provide scientists, engineers, and researchers in both academia and industry, who are interested in experimental and modeling research in microtechnology, with an overview and discussion of models and modeling techniques related to microfabrication systems. Models are compared to experimental observations when available. Models discussed in the book are useful for experimental or combined experimental and modeling studies in that they help in designing experiments, in interpreting experimental results, and in controling and optimizing a microfabrication system. They enable rapid progress towards commercialization.

This book is divided into six chapters, each of which deals with models related to a microfabrication system. Chapter 1 discusses mathematical models for predicting the evolution of a surface being bombarded by ions from an ion beam. Chapter 2 presents mathematical models relevant to predicting heat profiles, heat build-up, thermal deformations, and microstructure accuracy related to x-ray lithography. In Chapter 3, we deal with mathematical models predicting the growth of a microobject through laser chemical vapor deposition. Chapter 4 discusses models for predicting growth and behavior of microstructures through laser photopolymerization. In Chapter 5, we discuss mathematical models relevant to the prediction of surface erosion from laser ablation. Chapter 6 deals with heat transport in thin films. Thin films are included here because of their importance in many types of manufacturing systems, including those for high aspect ratio.

We wish to thank our colleagues Terry McConathy for proof reading the manuscript, Mike Vasile for valuable comments on Chapter 1, Mats Boman for reading Chapters 3 and 4, Philip Coane for reading Chapter 2 of the manuscript, and Dave Meng for his help with Latex. Last but not least, we wish to thank our student Lixin Shen for his help in reproducing figures for the text.

Louisiana Tech University,   *Raja Nassar*
February 2003   *Weizhong Dai*

Contents

1
Ion Beam

1.1 Introduction

Ion beam technology is of vital importance in the area of microfabrication. Focused ion beam (FIB) technology has found numerous applications in experimental electronic and electro-optic device fabrication (Kalburge et al. 1997; Konig et al. 1998), ion implantation research (Musil et al. 1996; Crell et al. 1997), device patterning (Gamo 1991; Nagamuchi et al. 1996), and FIB simulated deposition of conductors and opaque films (Vasile and Harriott 1989; Gross et al. 1990). Commercial uses of FIB include preparation of transmission electron microscopy (TEM) specimens, integrated circuit (IC) design and repair , diagnostics, and reverse engineering in the IC industry (Nikawa 1991; Banerjee and Livengood 1993; Prewett 1993; Giannuzzi and Stevie 1999; Krueger 1999). The FIB technology is based on the formation and deflection of high resolution ion beams from a liquid ion source (Bell and Swanson 1985; Orloff 1993). New applications for FIB are in high aspect ratio microstructure fabrication. Exploration of FIB microfabrication at the laboratory scale has been undertaken by Ishitani (1990), and Young (1993). Vasile et al. (1999, 1998, 1991) demonstrated the potential application of FIB in microstructure fabrication by producing microtools for ultraprecision machining, microsurgical tools and manipulators, tips for scanning probe microscopy, and arbitrary three-dimensional shapes. One aspect of the FIB microfabrication capability is its usefulness in the making of embossing masters or molds for mass production of microstructures.

Ion beam technology would benefit significantly from improved mathematical modeling of the ion beam process coupled with versatile ion beam control software. The benefits would come from more efficient utilization of the process and increased accuracy. Optimization of the beam diameter and the pixel address scheme in FIB will save considerable time and improve the accuracy and resolution of existing applications. New applications like micromold fabrication or rapid prototyping on a microscale become possible with more versatile ion beam control. In FIB technology, all of the processes (whether material is deposited or removed) have their dimensional resolution limited by the ion intensity profile relative to the pixel size and dwell time at the boundaries. High aspect ratio in the desired geometry provide the next most important dimensional limiting factor. In this chapter we present models related to ion beam large area sputtering and to focused ion beam micromachining.

1.2 Large Area Sputter Modeling

Early work on the modeling and simulation of surface profile evolution was done mainly for sputtering with a large area, where the ion beam is uniform, spatially continuous, and much larger in its lateral dimensions than the features being ion milled. A line-segment model of the surface movement dominated by the ion beam angle of incidence was the most fruitful approach. The final geometry is predicted given the elapsed time, the local material removal rates, and the starting geometry.

Nobes et al. (1969) developed a model to predict the equilibrium topography of an amorphous solid surface sputtered by a uniform ion beam. It was assumed that there was no redeposition of surface material and that surface changes occur only as a result of ejection of atoms with no surface diffusion. The sputter yield $S(\theta)$ is defined as atoms removed from the target per incident ion. The sputter yield is a function of the angle of incidence (angle formed between the ion beam at a target point on the surface and the normal to the surface at that point). As seen in Fig. (1.2.1), the value of $S(\theta)$ at $\theta = 0$ increases to a maximum at $\theta = \theta_p$ which then decreases to zero as θ approaches $\pi/2$. The exact form of $S(\theta)$ and the value of θ_p depend partially on the target material, ion type and energy, and the form of the energy deposition function. Based on these considerations, the rate of erosion of the surface in the normal direction for an isotropic target is expressed as

$$\phi S(\theta)/\eta \qquad (1.2.1)$$

where η is the atomic density defined as the number of atoms per unit volume of the target and ϕ is ion flux or ions/sec striking a unit area of a surface at an angle θ to the surface normal.

Consider a section of a surface at time t in the plane of the first quadrant (xoy) that is exposed to a uniform ion flux in the y direction of magnitude ϕ ions/sec/unit length in the x direction. Let the surface in the plane be defined as $y = f(x,t)$ and consider two points A and B on the surface as depicted in Fig. (1.2.2). Denote the angle of incidence at A by θ and at B by $\theta + \delta\theta$. The rate of bombardment per unit length at A is $\phi \cos\theta$, and the rate of recession or erosion of the surface at A in the normal direction is seen to be $\phi S(\theta)\cos\theta/\eta$. Hence, the rate of erosion in the y direction is

$$-\frac{\partial y}{\partial t} = \phi S(\theta)/\eta. \qquad (1.2.2)$$

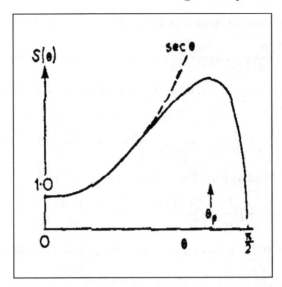

Fig. 1.2.1. Variation of Sputter yield S(θ) with the angle of incidence, θ (reprinted from Nobes et al. 1969 by permission of Kluwer Academic Publishers)

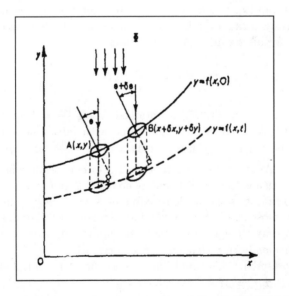

Fig. 1.2.2. Diagram of the geometry of erosion of a curve by ion beam bombardment in the y direction (reprinted from Nobes et al. 1969 by permission of Kluwer Academic Publishers)

Likewise, the rate of erosion at B in the y direction is

$$-\frac{\partial}{\partial t}(y + dy) = -\frac{\partial}{\partial t}(y + \delta x \frac{\partial y}{\partial x}) = \phi S(\theta + \delta \theta)/\eta. \qquad (1.2.3)$$

Here, δx and $\delta \theta$ imply dx and $d\theta$ in the limit. The first order Taylor expansion of Eq. (1.2.3) gives

$$-\frac{\partial}{\partial t}(y + \delta x \frac{\partial y}{\partial x}) = \frac{\phi}{\eta}(S(\theta) + \delta \theta \frac{dS(\theta)}{d\theta}). \qquad (1.2.4)$$

Subtracting Eq. (1.2.4) from Eq. (1.2.2) gives in the limit

$$\frac{\partial}{\partial t}(\frac{\partial y}{\partial x}) + \frac{1}{\delta x}\frac{\partial y}{\partial x}\frac{\partial}{\partial t}(\delta x) = -\frac{\phi}{\eta}\frac{dS(\theta)}{d\theta}\frac{d\theta}{dx}. \qquad (1.2.5)$$

When $\frac{\partial}{\partial t}(\delta x) = 0$, the recession is only in the y direction and is given by

$$\frac{\partial}{\partial t}(\frac{\partial y}{\partial x}) = -\frac{\phi}{\eta}\frac{dS(\theta)}{d\theta}\frac{d\theta}{dx}. \qquad (1.2.6)$$

Expression (1.2.6) applies when erosion of the surface does not proceed in the x direction. this situation occurs when part of the feature is parallel to the y direction of the beam. Equation (1.2.6) gives the time rate of change of the slope of the curve $y = f(x)$ as a function of the rate of change of the sputter yield with θ. From Eq. (1.2.6), one can determine the approach of the curve to its equilibrium form. Equilibrium occurs when the left-hand side of Eq. (1.2.6) is zero for all x values. At equilibrium,

$$\frac{\phi}{\eta}\frac{dS(\theta)}{d\theta}\frac{d\theta}{dx} = 0. \qquad (1.2.7)$$

One solution to Eq. (1.2.7) is obtained when $\frac{d\theta}{dx} = 0$ or θ is constant for all x values. As such, a continuous line with initial slope θ_0 will erode or recede in the negative y direction at a rate $\frac{\phi S(\theta_0)}{\eta}$ and will maintain a constant slope. Another solution to Eq. (1.2.7) can occur when $\frac{dS(\theta)}{d\theta} = 0$. This condition means that $S(\theta)$ is independent of θ and a curve of any initial slope will maintain its slope while receding at a constant rate in the negative y direction. In most cases, $S(\theta)$ is not independent of θ. For these cases, the surface topography changes over time with sputtering except in the case of a plane. Additionally, for elements of a curve where $\frac{dS(\theta)}{d\theta} = 0$, the slope will remain unaltered. This condition occurs where the initial slope is $\theta = 0$ or $\theta = \pm\theta_p$ ($S(\theta_p) = Max$) for which $\frac{dS(\theta)}{d\theta} = 0$, (Fig. 1.2.1). Consider a surface concave downwards that is bombarded by a uniform ion beam in the y direction. Let A be a point on the surface where $\theta = \theta_p$. To the left of point A on the surface θ is greater than θ_p and to the right of point A less than θ_p. It is seen then from Eq.(1.2.6) that $\frac{\partial}{\partial t}(\frac{\partial y}{\partial x})$ or the time rate of change in the slope of the surface is negative in the region to the left of point A ($\frac{d\theta}{dx} < 0$ and $\frac{dS(\theta)}{d\theta} < 0$, since $\theta > \theta_p$) and is positive to the right of A ($\frac{d\theta}{dx} < 0$ and $\frac{dS(\theta)}{d\theta} > 0$, since $\theta < \theta_p$). Hence, one would expect that at equilibrium the curve tends to a line with slope (angle to the x axis) $= \theta_p$. Thus, if a sphere is bombarded, one expects to obtain a conical shape at equilibrium. For a curve concave upwards with $\theta < \theta_p$ it is seen, using a similar argument, that $\frac{\partial}{\partial t}\left(\frac{\partial y}{\partial x}\right)$ is negative which implies that at equilibrium θ approaches 0 and the curve approaches a horizontal line. On the other hand, if $\theta > \theta_p$, then $\frac{\partial}{\partial t}\left(\frac{\partial y}{\partial x}\right)$ is positive which implies that at equilibrium θ approaches $\frac{\pi}{2}$ and the curve a vertical line. Hence, one would expect that a continuous surface, concave upwards with θ in the range 0 to $\pi/2$, would develop into a vertical step.

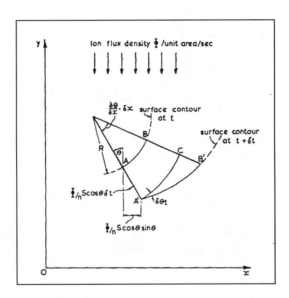

Fig. 1.2.3. Ersoion of a surface bombarded by a uniform ion beam (reprinted from Carter et al. 1971 by permission of Kluwer Academic publishers)

In a follow-up study, Carter et al. (1971) extended the present model to allow for surface motion or recession in the x direction. The surface in Fig. (1.2.3) depicts an eroding surface in the xoy plane with the ion beam incident in the y direction. In the figure, A and B are two adjacent points on the surface which erode in the time interval $t, t+\delta t$ to the points A', B'. AA' and BB' are perpendicular to the tangents at A and B. In time δt the surface erodes along AA' by a distance $\delta r = \frac{\Phi}{\eta}S(\theta)\cos(\theta)\delta t$. With $\Phi = \frac{\phi}{\cos\theta}$, the erosion rate, $\frac{\delta r}{\delta t}$ is as given by Eq. (1.2.1). The eroded distance BB' is $\frac{\Phi}{\eta}S(\theta+\delta\theta)\cos(\theta+\delta\theta)$. Taking the Taylor series expansion of BB' and subtracting AA' gives

$$CB' = \frac{\Phi}{\eta}\frac{\partial}{\partial\theta}(S\cos\theta)\frac{\partial\theta}{\partial x}\delta t \cdot \delta x \qquad (1.2.8)$$

It is seen also from Fig.(1.2.3) that

$$A'C \approx R\frac{\partial\theta}{\partial x}\delta x. \qquad (1.2.9)$$

The change in the tangential angle θ from A to A' is

$$-\delta\theta_t = \frac{CB'}{A'C} = \frac{\Phi}{\eta}\frac{\partial}{\partial\theta}(S\cos(\theta))\frac{\delta t}{R}. \qquad (1.2.10)$$

If θ is considered to be a function of x and t, then it is seen that

$$\delta\theta = \frac{\partial\theta}{\partial x}\delta x + \frac{\partial\theta}{\partial t}\delta t \text{ or } \frac{\delta\theta}{\delta t} = \frac{\partial\theta}{\partial x}\frac{\delta x}{\delta t} + \frac{\partial\theta}{\partial t} = \frac{1}{R\cos(\theta)}\frac{\delta x}{\delta t} + \frac{\partial\theta}{\partial t}. \qquad (1.2.11)$$

Hence, from Eqs. (1.2.10) and (1.2.11),

$$-\frac{\Phi}{\eta R}\frac{\partial}{\partial\theta}(S\cos(\theta)) = \frac{1}{R\cos(\theta)}\frac{\delta x}{\delta t} + \frac{\partial\theta}{\partial t}. \qquad (1.2.12)$$

From Fig.(1.2.3) it is seen that $\frac{\delta x}{AA'} = \sin\theta$. Hence, $\frac{\delta x}{\delta t} = \frac{\Phi}{\eta}S\cos(\theta)\sin(\theta)$. Substituting $\frac{\delta x}{\delta t}$ into Eq. (1.2.12) and replacing R by $1/\cos(\theta)\frac{\partial\theta}{\partial x}$ gives

$$\frac{\partial\theta}{\partial t} = -\frac{\Phi}{\eta}\cos^2\theta\frac{\partial S}{\partial\theta}\frac{\partial\theta}{\partial x} \text{ or } \frac{\partial\theta}{\partial t}\Big/\frac{\partial\theta}{\partial x} = -\frac{\Phi}{\eta}\cos^2\theta\frac{\partial S}{\partial\theta} \qquad (1.2.13)$$

The condition $\frac{\Phi}{\eta}\cos^2\theta\frac{\partial S}{\partial\theta} = $ constant represents contours at equilibrium in which a contour repeats itself upon sputtering, but moves in the x-direction. The condition $\frac{\Phi}{\eta}\cos^2\theta\frac{\partial S}{\partial\theta} = 0$ gives rise to stationary contours with movement restricted to the y-direction. This condition requires that either $\cos\theta = 0$ ($\theta = \pm\frac{\pi}{2}$) or $\frac{\partial S}{\partial\theta} = 0$ ($\theta = 0$, $\theta = \pm\theta_p$). Thus, equilibrium can be achieved only with vertical surfaces ($\theta = \pi/2$), with horizontal surfaces ($\theta = 0$) or with planes at an angle of inclination, $\theta = \pm\theta_p$. A combination of $\theta = 0$ and $\theta = \pi/2$ leads to a vertical step at equilibrium. Also, a combination of $\theta = \pi/2$ and $\theta = \theta_p$ leads at equilibrium to a three-dimensional surface in the form of a cylindrical cone or a cylindrical cone pit However, a combination of a horizontal plane and an inclined plane ($\theta = 0$ and θ_p) is not an equilibrium situation.

Catana et al. (1972) developed a computer program to simulate the development of a surface contour to equilibrium. Starting with the initial (x, y) coordinates of n equally spaced points on the x axis, the y-value at each point was reduced by an amount proportional to the $S(\theta)$ value at that point. The new slopes were then evaluated at each of the new points and the process repeated. It was noted that because a point on a curve moves along the surface normal, this method simulates movement in both the x and y directions.

Barber et al. (1973) showed how an earlier theory of Frank (1958) on the dissolution of crystals through chemical etching can be applied to predict the evolution of a surface bombarded with an ion beam. The method calls for plotting the reciprocal of the sputtering ratio $S(\theta)\cos(\theta)/S(0)$ (which is the normalized recession or erosion rate normal to the surface) on a polar graph paper (Fig. (1.2.4), from Carter et al. 1973). The resulting curve is called the erosion slowness curve. From this curve, trajectories are determined along which points of constant orientation move. This result is accomplished by first determining the normal to each point on the erosion slowness curve and allowing a point on the real surface with similar orientation to move along a trajectory in a direction parallel to this normal. The velocity with which a point moves along its trajectory is determined by the relative sputtering rate at that orientation. When two trajectories corresponding to different orientations meet, an edge is produced. The new direction of the edge movement is normal to the chord joining the two points where the original trajectories intersect the erosion slowness curve. From the array of trajectories that is constructed, one can derive the resulting surface topographies that evolve over time. The method is semi-quantitative and, as pointed out by the authors, its main source of error is in the determination of the slopes of the erosion slowness curve from which the normal and hence the erosion trajectories are derived. Predicted shapes starting initially with a sphere, a hemispherical trough, and a sinusoidal surface showed reasonably good qualitative agreement with experimental observations.

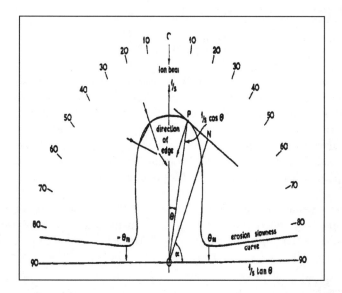

Fig. 1.2.4. Polar diagram of the erosion slowness curve (reprinted from Carter et al. 1973 with permission from Kluwer Academic Publishers)

Carter et al. (1973) showed how their theory fits into the framework of the theory by Frank (1958) as applied to the sputtering phenomena. In terms of Frank's kinematic theory, it is seen from Eq. (1.2.13) that

$$\frac{dx}{dt} = -\frac{\Phi}{\eta}\frac{dS}{d\theta}\cos^2\theta. \tag{1.2.14}$$

Equation (1.2.14) expresses the rate of motion of points of an orientation θ in the x-direction. An analogous expression is developed in the y direction given as

$$\frac{dy}{dt} = -\frac{\Phi}{\eta}(\frac{dS}{d\theta}\sin(\theta)\cos(\theta) - S). \tag{1.2.15}$$

The ratio of Eq. (1.2.14) to Eq. (1.2.15) gives the slope of the trajectories of points of constant orientation or θ value, namely

$$\frac{dy}{dx} = \frac{\frac{dS}{d\theta}\sin(\theta)\cos(\theta) - S}{\frac{dS}{d\theta}\cos^2\theta}. \tag{1.2.16}$$

The polar plot of the reciprocal of $\frac{S(0)}{S(\theta)\cos\theta}$ is equivalent to a Cartesian plot of $S(0)/S(\theta)$ as a function of $(S(0)/S(\theta))\tan\theta$. The normal to this curve has slope $\frac{d(\tan\theta/S(\theta))}{d(1/S(\theta))}$ which can be shown to be equal in absolute value to the slope of the normal to the erosion slowness curve. This result confirms that the trajectory of a point is parallel to the normal to the erosion slowness curve. Spatial velocity of points on a trajectory at a given orientation can be expressed from Eq. (1.2.14) and Eq. (1.2.15) as

$$v = \left[(\frac{dx}{dt})^2 + (\frac{dy}{dt})^2\right]^{1/2}, \tag{1.2.17}$$

which is useful in obtaining the progression of the eroding surface over time. From geometric considerations in Fig. (1.2.4), it is seen that $ON = OP\cos(90 - \alpha - \theta)$, $\tan\alpha = \frac{dy}{dx}$, and $OP = \frac{1}{S\cos\theta}$. From these relations, one can show that ON, representing the normal from the origin of the erosion slowness curve to the tangent at the point P, is proportional to the reciprocal of the spacial velocity v. Thus, the normal ON defines the direction of motion of a point with orientation θ and its reciprocal defines the velocity of the movement.

The same results, namely Eqs. (1.2.14) and (1.2.15) of Carter et al. (1973) were obtained by Ducommun et al. (1974), using different analytical techniques. Consider the initial surface $y = f(x, 0)$ to be the envelope of a family of straight lines defined as

$$\frac{y - f(x_i)}{x - x_i} = f'(x_i) \quad \text{or} \quad y - f(x_i) - xf'(x_i) + x_if'(x_i) = 0 \tag{1.2.18}$$

where $y_i = f(x_i)$, x_i is a point on the surface and $f'(x_i) = \tan\theta$. If we let $A(x_i) = \frac{\phi}{\eta}tS(\theta)$ represent the displacement (because of ion erosion) of a straight line parallel to itself, then the new family of straight lines transformed from (1.2.18) becomes

$$y - f(x_i) - xf'(x_i) + x_if'(x_i) + A(x_i) = 0. \tag{1.2.19}$$

The new surface is obtained from the solution of equation (1.2.19) and its derivative with regard to x_i, namely

$$-xf''(x_i) + x_if''(x_i) + \frac{\partial A(x_i)}{\partial x_i} = 0. \tag{1.2.20}$$

Replacing $A(x_i)$ and $f'(x_i)$ in (1.2.19) and (1.2.20) by their values gives

$$x = x_i + \frac{\phi}{\eta}t\cos^2\theta\frac{dS(\theta)}{d\theta}, \tag{1.2.21}$$

and

$$y = f(x_i) + \frac{\phi}{\eta}t[\sin\theta\cos\theta\frac{dS(\theta)}{d\theta} - S(\theta)]. \tag{1.2.22}$$

These equations are equivalent to Eqs. (1.2.14) and (1.2.15). Ducommum devised a computer program to solve eqs. (1.2.21) and (1.2.22) in order to follow the development of the surface contour and applied their results to a surface contour represented by $y = a\sin(x)$.

The models discussed so far apply to two-dimensional surfaces and a uniform ion flux. An analytical treatment for three dimensions using the method of characteristics is given by Smith and Wall (1980). The treatment allows for uniform and non-uniform ion distribution for situations where the beam is larger in area than the device being manufactured. Let the surface be denoted by the expression $A(\mathbf{r}, t) = 0$, where $\mathbf{r}$ is the position vector of a point on the surface. The equation describing the erosion rate in the normal direction is

$$\frac{\partial r_n}{\partial t} = -\frac{\phi}{N}S(\theta)\cos\theta. \tag{1.2.23}$$

Where r_n is the distance in the normal direction. At time $t + \delta t$, the new surface may be represented by $A(\mathbf{r} + \delta\mathbf{r}, t + \delta t) = 0$. It is seen that the point $(\mathbf{r} + \delta\mathbf{r}, t + \delta t)$ is at a distance

$$\delta\mathbf{r} = -\frac{\phi}{N}S(\theta)\cos\theta\delta t\frac{\nabla A}{|\nabla A|}, \tag{1.2.24a}$$

in the direction of the surface normal ($\boldsymbol{\eta} = \frac{\nabla A}{|\nabla A|}$), from the point $(\mathbf{r}, t)$ on the original surface, $A(\mathbf{r}, t) = 0$. Expanding $A(\mathbf{r} + \delta\mathbf{r}, t + \delta t) = 0$ using Taylor series gives $A(\mathbf{r} + \delta\mathbf{r}, t + \delta t) = A(\mathbf{r}, t) + \frac{\partial A}{\partial t}\delta t + \nabla A \cdot \delta\mathbf{r} +$ remainder. Substituting $\delta\mathbf{r}$ from Eq. (1.2.24a) into the Taylor expansion, one obtains as $\delta t \to 0$ the equation describing the rate of change in the surface caused by erosion.

$$\frac{\partial A}{\partial t} = \frac{\phi}{N}S(\theta)\cos\theta|\nabla A|. \tag{1.2.24b}$$

Equation (1.2.24b) may be squared and written in the alternative form

$$A_t^2 = \frac{\phi^2}{N^2} S^2 \cos^2 \theta \cdot (A_x^2 + A_y^2 + A_z^2). \qquad (1.2.24c)$$

Here the subscripts denote partial derivatives. Equation (1.2.24c) is a non-linear first-order partial differential equation whose solution may be obtained by the method of characteristics (Whitham, 1974). If one expresses the surface equation as $A(\mathbf{r}, t) = t - \sigma(\mathbf{r}) = 0$, where t is an explicit function of r, then Eq. (1.2.24c) becomes

$$1 - \frac{\phi^2}{N^2} S^2 \cos^2 \theta \cdot (\sigma_x^2 + \sigma_y^2 + \sigma_z^2) = 0, \qquad (1.2.25)$$

and $\nabla A = -(\sigma_x, \sigma_y, \sigma_z)$. If we denote $\mathbf{k}$ as the unit vector representing the ion beam in the z direction, then

$$\cos \theta = \mathbf{k} \cdot \boldsymbol{\eta} = -\frac{\sigma_z}{(\sigma_x^2 + \sigma_y^2 + \sigma_z^2)^{1/2}}. \qquad (1.2.26)$$

Substituting Eq. (1.2.26) into Eq. (1.2.25) leads to the equation

$$1 - \frac{\phi^2}{N^2} S^2 \sigma_z^2 = 0, \qquad (1.2.27)$$

This equation may be solved using the method of characteristics which transfers the first-order partial differential equation into a set of ordinary differential equations. The characteristic lines are given as follows:

$$\frac{dx}{dt} = -\frac{S'}{S} \cos^2 \theta \frac{\sigma_x}{\sigma_z (\sigma_x^2 + \sigma_y^2)^{1/2}}, \qquad (1.2.28)$$

$$\frac{dy}{dt} = -\frac{S'}{S} \cos^2 \theta \frac{\sigma_y}{\sigma_z (\sigma_x^2 + \sigma_y^2)^{1/2}}, \qquad (1.2.29)$$

$$\frac{dz}{dt} = \frac{1}{\sigma_z} + \frac{S'}{S} \frac{(\sigma_x^2 + \sigma_y^2)^{1/2}}{(\sigma_x^2 + \sigma_y^2 + \sigma_z^2)}, \qquad (1.2.30)$$

with the characteristic differential equations expressed as

$$\frac{\phi_x}{\phi} = \frac{d\sigma_x}{dt}; \quad \frac{\phi_y}{\phi} = \frac{d\sigma_y}{dt}; \quad \frac{d\sigma_z}{dt} = 0. \qquad (1.2.31)$$

In the two-dimensional case ($\sigma_y = 0$), Eqs. (1.2.28) and (1.2.30), with σ_x and σ_z derived from Eqs.(1.2.25) and (1.2.27), reduce to the same equations in Carter et al. (1979). When the beam is uniform, ϕ is constant over the (x, y) region. In this case, Eqs. (1.2.28) to (1.2.30) (each with a constant right-hand side) define straight line trajectories which determine the evolution of the surface topography. The rate of change in the shape of the eroding surface is determined by the velocity, $v = \{(\frac{dx}{dt})^2 + (\frac{dy}{dt})^2 + (\frac{dz}{dt})^2\}^{1/2}$, of motion of points along a characteristic. The ultimate equilibrium shape of the surface subjected to a uniform ion bombardment is a flat plane. In the non-uniform ion beam case, the characteristics will not be straight lines and Eq. (1.2.31) along with the characteristic lines must be solved numerically to determine surface evolution. Using a Kutta-Merson first-order method (Gerald 1978), a numeric solution was obtained to determine the development of the topography of a three-dimensional surface (elliptical hummock) under a uniform and Gaussian ion beam profiles (Figs. (1.2.5) and (1.2.6)). The sputter yield was given by $S(\theta) = 18.73845 \cos\theta - 64.65996 \cos^2\theta + 145.19902 \cos^3\theta - 206.04493 cos^4\theta + 147.31778 \cos^5\theta - 39.89993 \cos^6\theta$ which was obtained from experimental results for silicon. It is seen from Fig. (1.2.5) that for a uniform profile there is an overall change in the shape of the hummock as the ion dose increases. The ultimate shape at equilibrium is expected to be a horizontal plane. By contrast, the effect of the Gaussian beam is to produce initially a flattened surface, Fig.(1.2.6a), followed by a pit in the surface, Fig.(1.2.6b), which leads to a crater, Fig.(1.2.6c). In the case of the Gaussian beam, the effective beam width was of the same order of magnitude as the size of the Hummock. These results illustrate the importance of beam profile and beam width to the spacial resolution in surface erosion.

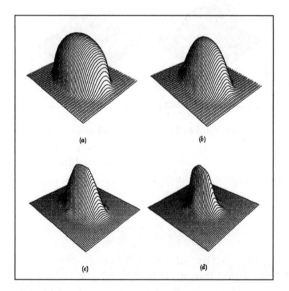

Fig. 1.2.5. Shape development by surface erosion of an amorphous elliptical hammock bonbarded by a unifrom ion beam with doses of **(a)** $\phi t/N = 0$, original shape, **(b)** $\phi t/N = 0.2$, **(c)** $\phi t/N = 0.4$, **(d)** $\phi t/N = 0.6$ (reprinted from Smith and Walls 1980 by permission of Taylor & Francis Ltd.
http://www.tandf.co.uk/journals)

In the method of characteristics, it is possible to obtain solutions that are physically unrealistic which can occur where the characteristic lines intersect forming points of discontinuity of the surface gradient. Such points are usually removed when carrying out the calculations. A model proposed by Hamaguchi et al. (1993) avoids this problem associated with discontinuity. This model, however, differs from previous models in that it predicts the evolution of a discontinuity point (nodal point) where the surface gradient changes abruptly. This result is accomplished by imposing a condition that identifies the physically meaningful solution of the partial differential equation describing the surface evolution. Hamaguchi derived a surface evolution partial differential equation (PDE) in the form

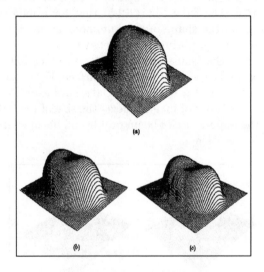

Fig. 1.2.6. Shape development by surface erosion of an amorphous elliptical hammock bombarded by a non-uniform, Gaussian ion beam with doses of (**a**) $\phi t/N = 0.2$, (**b**) $\phi t/N = 0.4$, (**c**) $\phi t/N = 0.6$ (reprinted from Smith and Walls 1980 by permission of Taylor & Francis Ltd. http://www.tandf.co.uk/journals)

$$u_t + c\sqrt{1 + u_x^2} = 0 \qquad (1.2.32)$$

where $c = c(t, x, y; u_x)$ is the surface etch rate in the normal direction, $y = u = u(x, t)$ defines the curve at time t, and the subscripts denote partial derivatives. A set of characteristic equations of the PDE for predicting the surface evolution (where the surface remains smooth and is described by the curve $y = u(x, t)$) under reactive-ion etching was obtained as

$$\frac{dx}{dt} = f_p = c_p\sqrt{1 + p^2} + c\frac{p}{\sqrt{1 + p^2}}, \qquad (1.2.33)$$

$$\frac{dy}{dt} = pf_p - f = c_p p \sqrt{1 + p^2} - c\frac{1}{\sqrt{1 + p^2}}, \tag{1.2.34}$$

$$\frac{dp}{dt} = -f_x - pf_u = -(c_x + pc_u)\sqrt{1 + p^2}, \tag{1.2.35}$$

where a subscript denotes a partial derivative (e.g., $f_p = \frac{\partial f}{\partial p}$), $p = u_x$ and $f(t, x, y, p) = c\sqrt{1 + p^2}$. The initial value for the equations is a point (x_0, y_0, p_0) on the initial curve $y = u(x, 0)$, where $p_0 = u_x(x_0, 0)$, and $y_0 = u(x_0, 0)$. The above equations describe surface evolution when the surface remains smooth with no discontinuities or shocks. When there are slope discontinuities, two equations were derived that predict the x and y components of the velocity of the slope discontinuity. These are expressed as

$$\frac{dX(t)}{dt} = \frac{f_l - f_r}{p_l - p_r}, \tag{1.2.36}$$

$$\frac{dY(t)}{dt} = \frac{p_r f_l - p_l f_r}{p_l - p_r}, \tag{1.2.37}$$

where $f_l = f(t, X(t), Y(t), p_l)$, $f_r = f(t, X(t), Y(t), p_r)$, $p_l = u_x(X(t) - 0, t)$, and $p_r = u_x(X(t) + 0, t)$. A condition was given as a criterion for identifying the solution to Eq. (1.2.32) that has physical significance. Let $u(x, t)$ be a continuous function that satisfies Eq. (1.2.32) except at points of discontinuity. The function $u(x, t)$ is then chosen as a solution if and only if $p = u_x$ around the discontinuities satisfies the condition

$$\frac{f_r - f}{p_r - p} \leq \frac{f_l - f_r}{p_l - p_r} \leq \frac{f_l - f}{p_l - p} \tag{1.2.38}$$

Using Eq. (1.2.38) and the shape of the curve (convex, concave, or neither convex or concave), Hamaguchi showed how to determine if a point discontinuity (i.e., a sharp corner) on the curve will evolve into a smooth arc or will propagate as a sharp corner or evolve as two sharp corners. Examples depicting the surface evolution of a semicircular trench, a rectangular trench, and an initial flat surface resulting from ion beam etching were presented. Similar expressions to Eqs. (1.2.36) and (1.2.38) have been given by Ross (1988) where he considered a similar problem.

Analytic methods dealing with characteristics have been discussed by Carter and Nobes (1984), and Smith et al. (1986). A recent review on the physics and applications of large area ion beam erosion is given by Carter (2001). In it he considers models to explain fine scale surface structures (less than 1 μm length scale) which cannot be explained by the above deterministic approach. A general nonlinear equation, Eq. (12) in Carter (2001), describing the evolution of local surface height is discussed. It includes, in addition to the deterministic surface erosion rate component, contributions from first and higher-order spatial derivatives of surface height and deterministic and random ballistic flux of surface atoms. Models dealing with simplified versions of the equation are discussed.

1.3 Focused Ion Beam

Focused ion beam (FIB) is a new technique that is promising in the manufacturing of three-dimensional microstructures of different shapes and depths. The ion beam is usually scanned on a grid surface (made up of pixels) along straight lines with constant or variable speed in the x and y directions. Modeling of focused ion beam sputtering is a different phenomenon from modeling of a large area ion beam on a masked substrate, even though the material removal process is the same. In FIB micromachining, there is no mask, and the geometric pattern is formed by ion beam deflection boundaries. The side wall profile in FIB milling depends upon the ion beam intensity profile and on the angle of incidence. Because of scanning, the total ion dose at a pixel is delivered in multiple exposures. The relationship between the beam width, the pixel size, and the time interval until pixel revisitation is critical for FIB etching.

Crow et al. (1988) described a simple model utilizing what is called vector scanning (that is a non-uniform scan rate) of the ion beam in order to etch 3-D microstructures with a desired profile. The model does not take into effect redeposition, ion beam reflection, or the dependence of sputter yield on the angle of incidence. The authors considered only 3-D profiles where the depth was a function of x, but constant in the y direction. The depth $z(x)$ is proportional to the ion dose $D(x)$, the sputter yield $S(\theta)$ which is a function of the angle of incidence θ, and the density of the material being etched, N. As such, it is seen that

$$z(x) \propto \frac{S(\theta)D(x)}{N}. \qquad (1.3.1)$$

With a constant beam current and spot size over time, it is seen that the ion dose is also proportional to the pixel dwell time, $t(x)$. Hence, Eq. (1.3.1) may be expressed as

$$z(x) \propto \frac{S(\theta)t(x)}{N}. \qquad (1.3.2)$$

To simplify the model equation, $S(\theta)$ was assumed constant. Since the dwell time is inversely related to the scan speed dx/dt of the ion beam, one may express the scan speed (based on Eq. (1.3.2) with $S(\theta)$ constant) as a function of $z(x)$

$$\frac{dx}{dt} = \frac{c}{z(x)}, \qquad (1.3.3)$$

where c is a constant. From Eq. (1.3.3), one may calculate the scan speed for different surface contours $z(x)$. It is seen that for a linear slanted surface $z(x) = ax$, the solution to Eq. (1.3.3), relating beam position to time, is $x(t) = \sqrt{\frac{2ct}{a}}$. Also, for a sine wave surface $z(x) = a\sin(x)$, Eq. (1.3.3) gives $x(t) = \arccos \frac{ct}{a}$. From these results it is possible to produce digitally generated deflection voltages for the ion beam that will control its scan speed so as to create the two surfaces. Using this technique, linearly slanted and sinusoidal surface contours were experimentally micromachined. Experimental results agreed qualitatively with what was expected from the model.

Ximen et al. (1990) developed a 3-D model that controls the scanning speed along a curved path which allows for the micromachining of nonlinear 3-D structures of different shapes such as V-shaped mirrors, 45° turning mirrors, parabolic turning mirrors, and laser arrays. In this treatment, it is assumed that the ion beam has a normal or Gaussian distribution, the sputter yield is a constant not dependent on the angle of incidence, and there is no redeposition of sputtered material and no ion beam reflection. Therefore, the model is strictly applicable to shallow craters (aspect ratio less than 2:1) where the angle of incidence is not too variable and where redeposition can be ignored. Considering a Gaussian ion beam in two dimensions, the etched depth at a point (x, y) on the surface was represented by the equation

$$Z(x,y) = N \sum_{x_p} \sum_{y_p} \frac{K}{2\pi} \exp[-\frac{(x - x_p)^2 + (y - y_p)^2}{2\sigma^2}] \cdot \omega(x_p, y_p)dt \quad (1.3.4)$$

where K is the etch rate in μm per second , N is the number of beam scans, σ^2 is the variance of the Gaussian distribution, dt is the minimum attainable dwell time in a (x_p, y_p) pixel space, $\omega(x_p, y_p)$ is a weighting factor determining the dwell time in a pixel space per scan, and (x_p, y_p) is the maximum point of the Gaussian distribution where the beam is focused or centered. Since the ion beam can only be scanned along straight lines in the x and y directions, the scanning strategy was to determine the speed in the x and y directions that gives a desired speed along the given path function, $y = f(x)$. The path function determines the depth profile of the microstructure. The depth in turn is proportional to the ion beam pixel dwell time which is inversely related to the scan speed. Multiplying the left- and right-hand sides of Eq. (1.3.4) by $dx_p dy_p$, taking $\omega(x_p, y_p) = 1$, and replacing the double sum by a double integral, one obtains that the smallest volume dv etched per scan is

$$dv = \int_{-\infty}^{\infty} \int_{-\infty}^{\infty} \frac{K}{2\pi} \exp[-\frac{(x - x_p)^2 + (y - y_p)^2}{2\sigma^2}]dx_p dy_p dt = K\sigma^2 dt. \quad (1.3.5)$$

This dv volume is equated to $2\sigma z(s)ds$ where ds is a small increment of the arc length s along the path function, $g(x, y)$, and $z(s)$ is the etched depth per scan. Hence,

$$2\sigma z(s)ds = K\sigma^2 dt. \quad (1.3.6)$$

From Eq. (1.3.6), one obtains the scan speed along the arc

$$\frac{ds}{dt} = \frac{K\sigma}{2z(s)}. \quad (1.3.7)$$

Equation (1.3.7) determines the scan speed along the arc as a function of the desired depth $z(s)$ to be etched. Since the ion beam can only be scanned along straight lines in the x and y directions, one may express the desired scan speed along the arc, $\frac{ds}{dt}$, as a function of the actual scan speeds in the x and y directions. Therefore, it can be seen from calculus that for any path function $y = f(x)$

$$\frac{dx}{dt} = \frac{ds/dt}{\sqrt{1 + (f')^2}} = \frac{K\sigma}{2z(s)\sqrt{1 + (f')^2}} \qquad (1.3.8)$$

and

$$\frac{dy}{dt} = \frac{f' \, ds/dt}{\sqrt{1 + (f')^2}} = \frac{f' K\sigma}{2z(s)\sqrt{1 + (f')^2}}. \qquad (1.3.9)$$

From Eqs. (1.3.8) and (1.3.9), one can determine the scan speeds of the ion beam in the x and y directions to obtain the desired scan speed along the arc in order to etch a desired depth $z(s)$. This approach will determine a digital scan for micromachining a 3-D structure with desired depth and shape. Figure (1.3.1) is an illustration of a micromachined circular crater with top and inverted view for a constant scan speed in the x and y directions (Fig. 1.3.1a), a constant scan speed ds/dt along the arc (Fig. 1.3.1b), and a variable scan speed along the arc (Fig. 1.3.1c). The constant and variable scan speeds along the arc were necessary in order to produce craters with uniform and sinusoidal depth profiles, respectively. Digital scan strategies were determined for etching V- and parabolic-shaped mirrors. The experimental results were not in complete agreement with simulated results from the model. These results were attributed to the aspect ratio of the mirrors being quite higher than 2:1 aspect ratio. For high aspect ratios, the deficiency of the model lies in the fact that the sputter rate is assumed to be constant and not a function of the angle of incidence θ. It is desirable to develop a 3-D model taking into account the dependency of sputter yield on the angle of incidence.

A model for focused ion beam induced gas etching has been developed by Harriot (1993). The advantages of using a gas etchant include enhancement of the etch rate, reduced redeposition of sputtered materials particularly in high aspect ratio microstructures, and reduced implantation of the ions in the target material. The process operates by introducing the precursor gas into the vacuum chamber of the ion beam system in the pixel area where the beam is focused. The gas flux is applied at a constant rate while the ion beam is being scanned. A basic assumption in this modeling approach is that etching is limited by the surface coverage of the precursor gas. There is a certain probability that the etchant gas will stick to the surface. Also, gas on the surface will evaporate after a certain residence time. The ion beam acts like a catalyst by providing energy to the substrate atoms and the gas molecules on the surface for a chemical reaction. It also enhances removal of the etched products. The chemical reaction can be expressed as

$$S + m(G) \longrightarrow SG_m, \qquad (1.3.10)$$

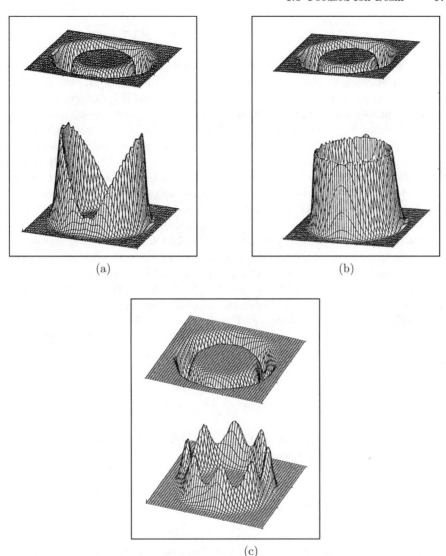

Fig. 1.3.1. Top and inverted views of a crater micromachined by applying (**a**)

constant x, y FIB scan speeds, (**b**) constant scan speed along the arc, and (**c**)
variable scan speed along the arc (reprinted from Ximen et al. 1990 with
permission from the American Vacuum Society)

where S represents the target material, and $m(G)$ represents the number of
gas molecules m needed to remove one atom from the substrate. Three factors
determine the etch gas coverage of the surface, namely surface adsorption,
evaporation, and removal by etching. As such, the rate of change in gas surface
coverage N is given by the ordinary differential equation

$$\frac{dN}{dt} = gF(\frac{N_0 - N}{N_0}) - \frac{N}{\tau_e} - msI(t)\frac{N}{N_0}, \qquad (1.3.11)$$

where N is the density of gas molecules on the surface, g is the probability of the gas sticking to the surface, F is the flux of the gas at the surface, N_0 is the density of empty or available adsorption sites on the surface, τ_e is the evaporation residence time, $I(t)$ is the ion beam flux which can be a function of time caused by beam scanning, and s is the number of reacted surface atoms that can be removed by a single ion. The etching process is achieved by scanning the ion beam, according to the pattern shown in Fig. (1.3.2), while the surface is exposed to the gas flux. The pattern is composed of contiguous pixels of square size. The beam is scanned repeatedly over the target area until the desired total ion dose or etched depth is achieved. As such, each pixel is exposed to the ion beam for a certain time t_p and then exposed again when the scan pattern is repeated. Etching occurs at a pixel site only during exposure. The time t_r between pixel exposures allows for additional adsorption of gas molecules on the site. For model application, it was assumed that $I(t)$ in Eq. (1.3.11) is constant and uniform over the pixel area during exposure time and is zero during the time between repeat exposures. Also, the pixel square dimension was considered to be equal to the radius of the beam. No allowance was made for overlap in beam intensity over adjacent pixels. Since $I(t) = 0$ during the time t between repeat exposures of any given pixel ($0 \leq t \leq t_r$), Eq. (1.3.11) reduces to

$$\frac{dN}{dt} = gF(\frac{N_0 - N}{N_0}) - \frac{N}{\tau_e}. \qquad (1.3.12)$$

The solution to this inhomogeneous ordinary differential equation can be obtained by standard techniques which give

$$N(t) = \frac{gF}{(\frac{1}{\tau_r} + \frac{1}{\tau_e})}\left\{1 - \exp[-(\frac{1}{\tau_r} + \frac{1}{\tau_e})t]\right\}; \qquad 0 \leq t \leq t_r, \qquad (1.3.13)$$

where $\tau_r = \frac{N_0}{gF}$.

If τ_e is much larger than τ_r, then $\frac{1}{\tau_e}$ may be ignored, and Eq. (1.3.13) reduces to

$$N(t) = N_0[1 - \exp(-\frac{t}{\tau_r})]. \qquad (1.3.14)$$

During pixel exposure, etching takes place, and the solution to Eq. (1.3.11) with $I(t) = I$ is readily obtainable to give

$$N(t) = \frac{gF}{(\frac{1}{\tau_r} + \frac{1}{\tau_e} + \frac{1}{\tau_p})} + c_1 \exp[-(\frac{1}{\tau_r} + \frac{1}{\tau_e} + \frac{1}{\tau_p})t]; \quad t_r \leq t \leq t_r + t_p, \quad (1.3.15)$$

where $\tau_p = \frac{N_0}{msI}$ and

$$c_1 = \frac{gF}{(\frac{1}{\tau_r} + \frac{1}{\tau_e})} \{ \frac{\frac{1}{\tau_p}}{\frac{1}{\tau_r} + \frac{1}{\tau_e} + \frac{1}{\tau_p}} - \exp[-(\frac{1}{\tau_r} + \frac{1}{\tau_e})t_r] \} \exp(\frac{1}{\tau_r} + \frac{1}{\tau_e} + \frac{1}{\tau_p})t_r.$$

(1.3.16)

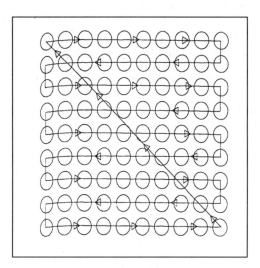

Fig. 1.3.2. Diagram illustrating the FIB raster scan or deflection pattern in the x, y directions (reprinted from Harriott 1993 with permission from the American Vacuum Society)

It is seen that the overall etch yield can be expressed in the form

$$Y = s(\frac{N(t_p)}{N_0}),$$

(1.3.17)

where $N(t_p)$ is evaluated from Eq. (1.3.15) at time $t = t_p$. This etch yield relation is justified since etching takes place only during pixel exposure to the ion beam and the ion beam flux is time homogeneous. One may simplify Eq. (1.3.17) under the assumption that $\tau_p \ll \tau_r \ll \tau_e$. In this case, Eq. (1.3.17) reduces to

$$Y = \frac{gF}{mI} - s\exp[-\frac{t}{\tau_p}].$$

(1.3.18)

If, in the operation of the ion beam, the pixel dwell time t_p is much larger than τ_p, then the total etch yield in Eq. (1.3.18) becomes only proportional to the ratio $\frac{F}{I}$. Experiments were conducted for etching SiO_2 and W using a 20keV Ga ion beam and XeF_2 etching gas. There was good agreement between model and experimental results for etch rate enhancement as a function of t_p or pixel dwell time.

An accurate, predictive mathematical model of the digitally driven focused ion beam sputtering process remains an obstacle to more general and efficient

application of focused ion beam technology. The most useful model would be
one which prescribes the ion beam dwell time at a particular pixel from the
geometry of the feature to be ion milled. In developing such a model, it is
important to take into account the dependence of sputter yield on the angle
of incidence. The sputter yield as a function of angle, $S(\theta)$, of monotonic solids
has been given by Yamamura et al. (1983) as the normal incidence multiplied
by functions which describe penetration as the incidence angle increases, and
particle reflection at high (grazing incidence) angles:

$$S(\theta_i) = Y(E)(\cos\theta_i)^{-f} \exp[-c((\cos\theta i)^{-1} - 1)], \qquad (1.3.19)$$

where c and f are functions of the physical constants of the ion solid combination. The expression $Y(E)$ is the sputter yield for ions incident on the
surface normal.

$$Y(E) = P\frac{S_n(E)}{1 + 0.35 U_s S_e(E)}[1 - (\frac{E_{th}}{E})^{1/2}]^{2.8}, \qquad (1.3.20)$$

where

$$P = 0.042\frac{E_L N}{R_L U_s}\alpha(M_2/M_1)Q(Z_2) \qquad (1.3.21)$$

and

$$S_n(E) = \frac{3.441\sqrt{E}\log(E + 2.718)}{1 + 6.355\sqrt{E} + E(-1.708 + 6.88\sqrt{E})}. \qquad (1.3.22)$$

where $\alpha(M_2/M_1)$, R_L, E_L, E_{th}, and $S_e(E)$ are calculated from expressions
in the physical constants; E is the kinetic energy of the projectile; U_s is the
cohesive energy of the solid (taken as the sublimation energy); and N is the
number of atoms/cm^3 in the target material. $Q(Z_2)$ must be given for each
target, from compilations in Yamamura, as though it were a physical constant. These semi-empirical formulae of Yamamura were chosen for their wide
applicability and accuracy in fitting experimental data on FIB milling. Figure
(1.3.3) shows the curve from Yamamura depicting the angular dependence of
the normalized sputter yield, $\frac{S(\theta)}{Y(E)}$, for Ga^+ on Si at 30 keV. The points are
experimental data. Yamamura's expression is consistent with the experimen-
tal data for all angles of incidence. The sputter yield reaches a maximum at
about 80° and then decreases very rapidly to zero as θ approaches 90 degrees.

Making use of the sputter yield by Yamamura, Nassar et al. (1998) developed a model for controlling the deflection pattern of a focused ion beam to
produce a microscopic structure with a predetermined geometry. The design
of the object is translated into a net of sputter depths as a function of position (pixel address) on a surface layout. The ion beam dwell time needed to
produce the sputter depth is computed as a function of material sputter yield,
angle of incidence, and intensity contributions from all other pixels(resulting
from ion beam distribution) in the address scheme. This procedure assumes
that the ion beam diameter is constant and larger than the pixel dimension.

The effect of redeposition is not considered in this model. Given an initial surface geometry, one must determine the dwell time for each pixel on the surface for each scan in the x and y directions, in order to produce a microstructure with a desired geometry. To demonstrate this concept, consider the parabolic cross-section in Fig.(1.3.4). In this case, the initial surface is given by the parabolic surface A. The desired surface to be etched is surface B. At any point (x_i, y_j) on surface A, there is a corresponding point (x_i, y_j) on surface B. The term ΔZ_{ij} represents the depth to be sputtered at that point. Considering all (x_i, y_j) points on surface A, one can generate surface B by sputtering a depth ΔZ_{ij} at each point (x_i, y_j). A distribution of the ion beam intensity implies that a point (x_i, y_j) on a surface will receive ion flux from another point (x_k, y_l) at which the beam is focused. With overlap in ion flux, one has to consider the surface in two dimensions as shown in Fig. (1.3.5). Because the angle of incidence θ changes as the etched surface evolves, it is necessary to use multiple scans for etching the final geometry. The three-dimensional structure will then be broken into a series of slices, each with thickness ΔZ_{ij} where each slice is sputtered in one scan. The depth ΔZ_{ij} sputtered at a point (x_i, y_j) can be expressed as

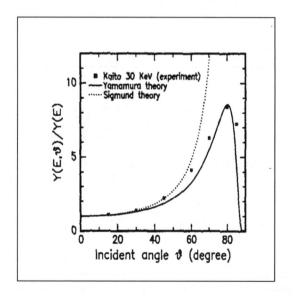

Fig. 1.3.3. Normalized sputtering yield of Ga$^+$/Si at 30 keV. The points represent experimental data (Kaito,personal communication). The solid line is from theoretical results by Yamamura 1983, and the dotted line from Sigmund's theory 1969 (reprinted from Vasile et al. 1999b with permission from the American Vacuum Society)

$$\Delta Z_{ij} = \int \int \frac{\phi(x,y)}{\eta} f_{x,y}(x_i, y_j) S(\theta_{x_i,y_j}) t_{xy} dx dy, \qquad (1.3.23)$$

where $\phi(x,y) =$ ion flux at point x, y in $cm^{-2}s^{-1}$, $\eta =$ atomic density of the target in atoms cm^{-3}, $S(\theta) =$ sputter yield in atoms per incident ion at point (x_i, y_j), $t_{xy} =$ dwell time of the ion beam at point (x, y) in seconds, and $f_{x,y} =$ ion beam density function in two dimensions.

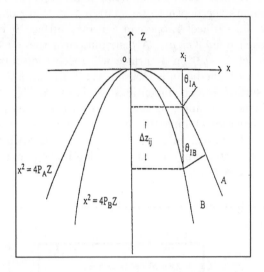

Fig. 1.3.4. A cross-sectional projection (at $y = 0$) of two parabolic surfaces A and B (reprinted from Nassar et al. 1998 with permission from the American Vacuum Society)

The sputter yield $S(\theta)$ is as given in Eq.(1.3.19). The angle of incidence, θ, at point (x, y) can be determined from the geometry of the surface. The beam distribution is considered to be the Gaussian bivariate density

$$f_{x,y}(x_i, y_j) = (\frac{1}{\sqrt{2\pi}\sigma})^2 \exp\left\{-[(x_i-x)^2 + (y_j-y)^2]/2\sigma^2\right\} \qquad (1.3.24)$$

In Eq. (1.3.24), $\phi(x,y)f_{x,y}(x_i, y_j)$ gives the ion flux contribution to point (x_i, y_j) from a beam focused at (x, y). The double integral gives the ion flux contribution at (x_i, y_j) from all points on the scanned surface. From Eq. (1.3.23), one may obtain a solution for the dwell time t_{xy} at every point (x, y) on the surface needed for etching an increment ΔZ_{ij} at point (x_i, y_j). Eq. (1.3.23) may be discretized to obtain

$$\Delta Z_{ij} = \sum_{k=1}^{n_1} \sum_{l=1}^{n_2} \frac{\phi(x_k, y_l)}{\eta} f_{x_k,y_l}(x_i, y_j) S(\theta_{x_i,y_j}) t_{kl} \Delta x_k \Delta y_l, \qquad (1.3.25)$$

where

$$f_{x_k,y_l}(x_i, y_j) = (\frac{1}{\sqrt{2\pi}\sigma})^2 \exp\left\{-[(x_i - x_k)^2 + (y_j - y_l)^2]/2\sigma^2\right\} \quad (1.3.26)$$

In this representation, the surface is divided into $n_1 n_2$ small square pixels, and the point (x_i, y_j) represents the midpoint of the ij pixel ($i = 1, 2, ..., n_1$; $j = 1, 2, ..., n_2$). If we denote

$$C_{k,l}(i,j) = \frac{\phi(x_k, y_l)}{\eta} f_{x_k,y_l}(x_i, y_j) S(\theta_{x_i,y_j}) \Delta x_k \Delta y_l, \quad (1.3.27)$$

then Eq. (1.3.25) may be expressed in matrix form as

$$\{C_{k,l}(i,j)\}\{t_{k,l}\} = \{\Delta Z_{ij}\}, \quad (1.3.28)$$

where $\{C_{k,l}(i,j)\}$ is an $n \times n$ matrix ($n = n_1 n_2$), $\{t_{k,l}\}$ is an $n \times l$ column vector, and $\{\Delta Z_{ij}\}$ is an $n \times l$ column vector. Equation (1.3.28) gives a set of n linear equations in n unknowns which was solved to give the dwell time $t_{k,l}$ at each pixel and hence the deflection or scan pattern of the ion beam needed for etching a depth ΔZ_{ij} at each ij pixel. After each scan, the angle of incidence θ at each point (x_i, y_j) is adjusted to correspond to the new surface generated. The pixel dwell time has been programmed into the ion beam deflection routine so that the ion beam scan leads to the creation of the desired geometry of the microstructure. It is clear also that from the dwell times $\{t_{k,l}\}$ one may predict from Eq. (1.3.25) the edge profile for pixels at the periphery of the scan region.

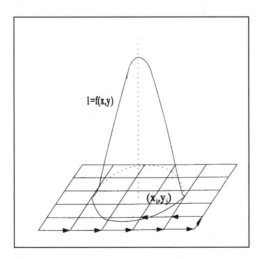

Fig. 1.3.5. An overlay of a Gaussian ion beam intensity profile on the pixel grid with scan direction indicated by arrows. The dose received at point (x_i, y_j) is a function of the ion beam distribution over the surface and the dwell time per pixel (reprinted from Nassar et al. 1998 with permission from the American Vacuum Society)

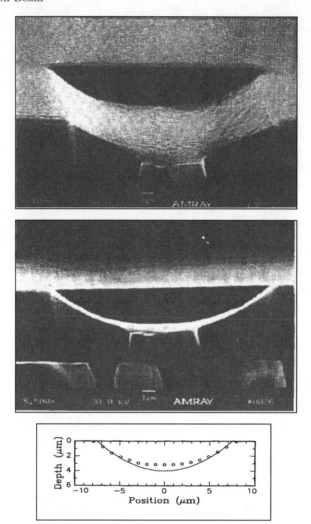

Fig. 1.3.6. SEM image of a parabolic trough (**a**) viewed at 75^0, (**b**) viewed at 90^0, and (**c**) comparison between the specified geometry and experimental data, scale bar =1 μm (reprinted from Vasile et al. 1999b with permission from the American Vacuum Society)

The algorithm for solving the linear system in Eq. (1.3.28) was incorporated into the ion beam deflection control system for model verification (Xie 1999). It is possible to reduce the storage requirement for the matrix $\{C_{k,l}(i,j)\}$ from $n \times n$ to η by storing only distinguishable element values in the matrix. Also, in carrying the computation, one need not include all pixels in the scanned region, but only those that have significant dose contribution to pixel ij. This approach allows one to obtain solutions for large systems in a reasonable time. Experimental verification of the model was accomplished by ion milling parabolic, hemispheric, and cosine troughs which are characterized by a plane of symmetry in the ion milled cavity (Vasile et al. 1999). The same geometric shapes were also ion milled using a rotational axis of symmetry to generate sinusoidal rings, and parabolic and hemispheric dishes. These shapes were chosen to test the capability of the model to ion mill complex structures under a wide range of incident angles. A Ga^+ ion beam operated at 20 keV with a current between 2.0 and 2.5 nA was used for the milling operation. The ion beam had a Gaussian distribution with a standard deviation $\sigma = 0.263\mu m$. Single crystal SI(001) was used for substrate material. The dimension scales of the SEM and ion beam had a common calibration to ion mill and measure the cavities. Figures (1.3.6)- (1.3.9) show SEM views of the ion milled cavities at various angles. Measurements made from a 90° view were compared to predicted values from the model. These figures show the capability of the model to control the deflection pattern of the ion beam for the milling of 3-D microstructures. The maximum depth measured for any of the cavities was between 0.8 and 0.9 of the expected depth from the model. This discrepancy was attributed to redeposition and slowly decreasing ion current over the milling time duration. Redeposition is no doubt of importance in the milling of high aspect ratio microstructures with steep walls and high depth at the center position, as is the case for the hemisphere which showed relatively strong redeposition effect. It is possible to incorporate a redeposition rate in the model that is a function of the (x, y, z) position in the cavity. At present, this approach does not seem feasible since there is a lack of empirical evidence as to what function to choose.

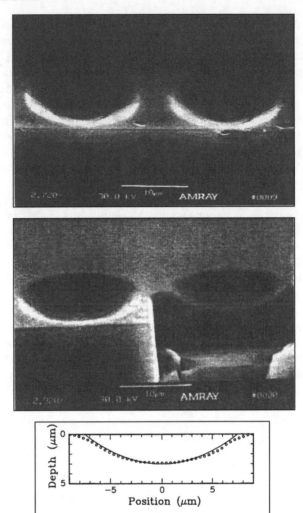

Fig. 1.3.7. (a) SEM image of a parabola with z-axis of symmetry viewed at 60^0, (b) image of the cross section of the same parabola viewed at 60^0, (c) comparison between the geometric specification (solid line) and experimental data, scale bar = 10μm (reprinted from Vasile et al. 1999b with permission from the American Vacuum Society)

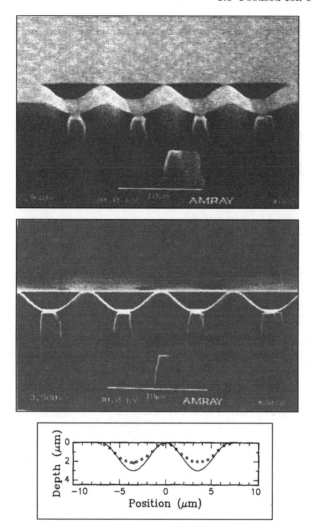

Fig. 1.3.8. SEM image of a sinusoidal surface (**a**) viewed at 75^0, (**b**) at 90^0, and (**c**) comparison between the middle two sinusoidal waves (solid line) and experimental data, Scale bar $= 10\mu$m (reprinted from Vasile et al. 1999b with permission from the American Vacuum Society)

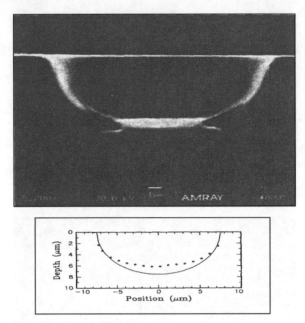

Fig. 1.3.9. (a) SEM image of a cross section of a hemisphere with rotational symmetry along the z-axis viewed at 90^0, (b) Comparisons between the geometric specification (solid line) and experimental data. Scale bar $= 1$ μm (reprinted from Vasile et al. 1999b with permission from the American Vacuum Society)

2

X-ray Lithography

2.1 Introduction

X-ray lithography is an important process used in the manufacture of high aspect ratio microstructures. It also represents a promising alternative to ultraviolet optical lithography in the manufacture of large scale integrated circuits with high resolution. The process involves the irradiation of a mask and a photoresist deposited on a substrate. A gas such as helium circulates in the gap between mask and resist as a cooling device. The mask is made of a thin membrane of low atomic weight material, through which x-rays can be transmitted readily, and a metallic absorbing pattern such as gold which has a high x-ray absorbing capability. The mask layer creates a desired pattern on the photoresist by selectively allowing the transmission of irradiation from an x-ray beam. Polymethylmethacrylate (PMMA) and other photoresists used in patterning microelectromechanical systems (MEMS) are poor conductors of heat compared to the supporting silicon substrate.

Exposure time to an x-ray source may limit the use of the technique for rapid manufacturing of microdevices needed for commercialization. Short exposure times may be achieved from high flux synchrotron sources or from intense pulsed plasma x-ray sources. Synchrotron radiation has a near-constant radiation power, and the maximum power on the mask is about 150 mW/cm^2. Radiation from a plasma source is emitted by a short-lived plasma, high in temperature and density, which is produced by an electrical discharge or a high power laser source. Plasma intensity is distributed isotropically and is delivered in short pulses of 1-100 ns in duration with power of 10^5-10^6 W/cm^2. With high flux or short pulse energy deposition, heating of the photoresist and mask membrane may develop rapidly. The heat can lead to distortions in the mask and in the exposed area of the resist compared to the unexposed area. Additionally, stresses between the resist and substrate may develop caused by differential heat conduction leading to bonding problems where the resist substrate bonding is weak.

For better understanding of the process and for the purpose of design and optimization, it is effective to consider heat transfer models that can predict temperature distributions and thermal stresses that can lead to distortions in the mask and photoresist and to loss of accuracy in the manufacturing process (Chaker et al. 1992; Ameel et al. 1994; Li et al. 1996; Heinrich et al 1983; Hegab et al. 1998). In this chapter, we present and discuss models on mask heating and distortion, on resist heating, and on the accuracy of structure transfer.

2.2 Mask Heating and Distortion

Temperature distributions and distortions caused by thermal stress in a boron-doped silicon mask exposed to intense synchrotron x-ray irradiation has been addressed by Heinrich et al. (1983). Heat distribution in the mask was assumed to be affected by heat conduction in the mask, heat conduction across the mask surface by means of the gas in the gap, and heat loss caused by radiation. The two-dimensional heat equation, used to represent the temperature distribution in the mask, is expressed as

$$cd\frac{\partial T}{\partial t} = kd(\frac{\partial^2 T}{\partial x^2} + \frac{\partial^2 T}{\partial y^2}) + \text{sources-losses} \tag{2.2.1}$$

where c is the heat capacity, k is thermal conductivity, and d is the thickness of the mask membrane. Heat loss from the surface exposed to irradiation caused by thermal radiation is expressed as

$$l_{rad} = \epsilon\sigma[(T + T_0)^4 - T_0^4] \tag{2.2.2}$$

where T_0 is the ambient temperature, T is the temperature difference between mask and T_0, ϵ is the emissivity, and σ is the Boltzmann constant. When T is small relative to T_0, Eq. (2.2.2) can be reduced to the linear form

$$l_{rad} = \alpha_{rad}T \tag{2.2.3}$$

where α_{rad} is the heat transfer coefficient caused by radiation and is equal to $4\epsilon\sigma T_0^3$. Heat loss from the same surface caused by convection is calculated as

$$l_{conv} = \alpha_{conv}T \tag{2.2.4}$$

where $\alpha_{conv} = 5.5 \times 10^{-4}(T/K)^{1/4} \ W/cm^2 K$ for helium gas at 1 bar pressure. Another heat loss source is due to heat flow across the mask surface to the gas in the gap. In the case of helium gas, this heat flow is determined as

$$l_{He} = \alpha_{He}T \tag{2.2.5}$$

where $\alpha_{He} = k_{He}/d_{He}$ and d_{He} is the proximity gap between mask and resist. For a proximity distance of 50 μm, the heat transfer coefficients were calculated to be

$$\alpha_{He} = 0.3W/cm^2K, \tag{2.2.6}$$

$$\alpha_{rad} = 0.0012W/cm^2K, \tag{2.2.7}$$

where $T_0 = 300K, \epsilon = 1$, and

$$\alpha_{conv} = 0.00055(T/K)^{1/4}W/cm^2K \tag{2.2.8}$$

It is seen that heat loss to the helium gas in the gap is the most important mechanism contributing to the cooling of the mask. If we let α_{eff} denote the effective heat transfer (loss) coefficient obtained as the sum of $\alpha_{He}, \alpha_{rad}$, and α_{conv}, we may rewrite Eq. (2.2.1), upon dividing both sides by α_{eff}, in the form

$$\tau\frac{\partial T}{\partial t} = \alpha^2(\frac{\partial^2 T}{\partial x^2} + \frac{\partial^2 T}{\partial y^2}) + T\max - T \tag{2.2.9}$$

where $\tau = cd/\alpha_{eff}$ is referred to as the thermal rise time constant, $\alpha = \sqrt{kd/\alpha_{eff}}$ is the thermal diffusion length, and $T\max = P/\alpha_{eff}$ is the maximum temperature rise at the point where the absorbed power intensity P (source) is maximum.

For a homogeneous power intensity distribution in the mask and under steady-state conditions, Eq. (2.2.9) reduces to $T(x,y) = T_{max} = P/\alpha_{eff}$ which implies that the maximum temperature rise is everywhere in the mask membrane. This maximum temperature is 0.33K for a 2 μm thick silicon membrane and gap width of 50 μm. For the case of steady-state conditions and an inhomogeneous power distribution in space, namely $P(x,y) = P_0$ for the region $-b/2 \leq x \leq b/2$ (b is the width of the irradiation region) and $P(x,y) = 0$ elsewhere, Eq.(2.2.9) reduces to a one- dimensional equation

$$\alpha^2(\frac{\partial^2 T}{\partial x^2}) + T\max - T = 0 \tag{2.2.10}$$

which gives the solution

$$T(x,y) = (P_0/\alpha_{eff})[1 - \exp(-b/2\alpha)\cosh(x/\alpha)] \tag{2.2.11}$$

for $-b/2 \leq x \leq b/2$, and

$$T(x,y) = (P_0/\alpha_{eff})\sinh(b/2\alpha)e^{-|x|/\alpha} \tag{2.2.12}$$

for $|x| \geq b/\alpha$. Here, the maximum temperature occurs at $x = 0$, or

$$T(0,y) = (P_0/\alpha_{eff})[1 - \exp(-b/2\alpha)]. \tag{2.2.13}$$

If b is large compared to α, the temperature is the same in the irradiation region and is $T(0,y) = T_{max}$ or P_0/α_{eff} which is about 0.33K.

For a homogeneous power intensity and an unsteady-state case with initial temperature zero, the solution to Eq. (2.2.9) gives

$$T(x, y, t) = T_{\max}[1 - \exp(-t/\tau)] \tag{2.2.14}$$

with $T_{\max} = 0.33K$.

In the case of irradiation by a plasma source, it was argued that, because the duration of the high intensity plasma pulse (about 20 ns) is much shorter than the time needed for heat transport through the silicon mask or through the helium gas, severe heating of the mask can occur. It was predicted that, resulting from the high plasma irradiation power of 10^5-10^6 W/cm^2, the temperature in the mask will rise during the 20 ns pulse by 6-60 K inside the gold absorber and by 2.5-25 K in the silicon membrane of the mask, depending on the power intensity. The temperature will be expected to decrease exponentially between pulses.

In order to determine mask distortion caused by temperature rise, one can relate the displacement (distortion) u in the one-dimensional case to the elastic deformation or strain (ϵ) and to the stress σ based on the differential equation

$$du/dx = \epsilon + aT \tag{2.2.15}$$

where aT is the thermal expansion and $\epsilon = \frac{1-v}{E}(\sigma - \sigma_0)$. In the radial case with symmetry, the following equations are applicable:

$$du/dr = \epsilon_r + aT,$$

$$\sigma = \frac{E}{1 - v^2}(\epsilon_r - v\epsilon_t) + \sigma_0,$$

$$u/r = \epsilon_r + aT,$$

$$\sigma_t = \frac{E}{1 - v^2}(\epsilon_t - v\epsilon_r) + \sigma_0,$$

and

$$(\partial/\partial r)/(r\sigma_r) - \sigma_t = 0. \tag{2.2.16}$$

Here, the index r is radial and t is tangential, E is the Young's modulus, v is the Poisson ratio (assumed the same as for pure silicon), and σ_0 is the original stress introduced by boron doping.

Considering a one-dimensional case, it is seen that the solution to Eq. (2.2.15) depends on the temperature as a function of x, $T(x)$. Calculations based on temperature distributions Gaussian in shape (representative of the synchrotron case) showed that the maximum displacement was less than 0.03 μm under the assumption of a maximum temperature of 1 K and a Gaussian distribution width of 0.5 cm. For the case of radial symmetry and a homogeneous temperature distribution on an area of the mask circular in shape, it was found from the analytical solution of Eq. (2.2.16) that the maximum displacement was about 0.015 μm for the case of a circular area 2 in. in diameter and a maximum temperature rise of 1 K. It was determined that, even with plasma irradiation, the displacement can be expected to remain within acceptable limits provided that irradiation of the mask was homogeneous. The one-dimensional model based on simple assumptions of equilibrium of stretching forces and zero thermal expansion of the silicon rim was used to predict the effect of gold coverage on the silicon membrane. It was concluded that the maximum distortion or displacement with gold coverage (with absorber pattern equally distributed) was less than twice that which occurred in the non-covered membrane.

Vladimirsky et al. (1989) performed experimental measurements of the temperature rise for synchrotron exposure by using eight 50 nm thick gold resistors fabricated near the center of a boron-doped Si membrane 2.5 μm in thickness. The resistors were placed on a polyimide insulating layer (1.5 μm thick) and as such were not in contact with the Si membrane. The sensitivity of the measuring device was of the order of $.02^0 C$. The mask-substrate or wafer assembly was scanned through a synchrotron x-ray beam at the rate of 0.01 cm/sec. The temperature rise was recorded for 25 mm and 35 μm gap settings both in vacuum and in a 20 Torr pressure helium atmosphere. Temperature measurements were compared to the maximum temperature rise, P_0/α_{eff}, from Eq. (2.2.13). The maximum temperature rise measured was 25 $^0 C$ for both 25 mm and 35 mm gaps in vacuum and those calculated were 6.7 and 4.1, respectively. Also, the maximum temperature rise measurements for 20 Torr He were 13 $^0 C$ and .07 $^0 C$ as compared to 2.5 and 0.3 calculated. This large discrepancy between observed and calculated temperature rise may be explained, according to the authors, assuming different parameters in P_0/α_{eff}. However, it is not clear how much influence the polyimide insulating layer had on these temperature measurements.

With plasma irradiation, the mask is subjected to repetitive short x-ray pulses. Most of the fluence is transmitted through the mask membrane to the resist, but a significant fraction can still be absorbed by the membrane. However, the gold absorber pattern on the mask membrane absorbs most of the fluence. The maximum temperature rise at the interface between absorber and membrane in the case of instantaneous deposition can be expressed as

$$\Delta T_{\max} = \frac{J_0}{\rho c \lambda} \tag{2.2.17}$$

where J_0 is the fluence transmitted through the membrane to the resist (Ballantyne and Hyman 1985). The temperature distribution in the mask membrane is governed by the heat equation

$$\frac{\partial T}{\partial t} = a\frac{\partial^2 T}{\partial x^2} + \frac{\phi e^{-x/\lambda}}{\rho c \lambda} \tag{2.2.18}$$

where a is the thermal diffusivity and $\frac{1}{\lambda}\phi e^{-x/\lambda}$ is the flux at depth x. An analytic solution using Laplace transform can be obtained under certain conditions where the membrane is considered semi-infinite in length (valid for short pulse length). A solution for the temperature rise at the interface is given by

$$T_{int} = \Delta T_{max}\{(\frac{e^\beta - 1}{\beta}) - \frac{e^\beta}{2\beta}[\text{erf } c(-\sqrt{\beta}) \tag{2.2.19}$$

$$+\frac{1}{\epsilon^2}\text{erf } c(2\gamma - \sqrt{\beta}) - \text{erf } c(\gamma - \sqrt{\beta}) + \text{erf } c(\gamma + \beta)]$$

$$+\frac{h}{\lambda\epsilon\beta}\text{erf } c(\gamma) - \frac{1}{2\beta}[(\frac{2h}{\lambda} - 1)\text{erf } c(2\beta) - 1]$$

$$+\frac{1}{\sqrt{\beta\pi}}(1 + e^{-4\gamma^2} - \frac{2}{\epsilon}e^{-\gamma^2})\}$$

where $\beta = a\tau/\lambda^2, \gamma = h/(2\sqrt{a\tau})$, h is the absorber thickness, τ is the pulse duration, and $\epsilon = e^{h/\lambda}$. Figure (2.2.1) presents the normalized temperature rise, $T_{int}/\Delta T_{max}$, as a function of x-ray pulse duration (*secs*) for different x-ray wavelengths, and $\epsilon = 35.7$. It is seen that the temperature rise decreases with an increase in pulse duration. Thermal stress can be expected to occur at the interface between absorber and membrane where the temperature gradient is large. Under the assumptions that the temperature rise in the absorber is uniform, the membrane is rigid, and stress is two-dimensional, the shear stress at the interface is expressed as

$$\tau_{xz}(x,0) = \frac{14.28Ea\Delta T}{1 + 1.2\phi_r} \cdot \frac{\sinh(kL/2)\sinh(kx)}{kL + \sinh(kL)} \tag{2.2.20}$$

where $\phi_r = \frac{\sinh(kL)-kL}{\sinh(kL)+kL}$, $k = \sqrt{\frac{1-2v}{2(1-v)}} \cdot \frac{\pi}{2h}$, L is the length, h the height of the absorber, and the numerical constants are defined for gold with $v = 0.42$. A plot of $\tau_{xz}/Ea\Delta T$ against (L/h) shows that the shear stress tends to a limit as L/h increases in value. The shear stress in the limit is given by $\tau_{xz}(L/2,0)$ $= 3.25Ea\Delta T$ and the axial compressive stress is expressed approximately as $\sigma_{xx}(x,0) \approx -\frac{Ea\Delta T}{1-v} \approx -1.7Ea\Delta T$. Shear stress dominates over compressive stress at the interface for $L/h > 2$. Both shear and compressive stresses were highest at the corner point where the edge of the absorber block meets the membrane. It was anticipated that at that point shear stress would lead to failure in the form of delamination. Failure may occur as a result of a single-pulse irradiation or by fatigue resulting from repetitive pulses. However, the limits for such failures to occur remain speculative. For gold, the yield stress was taken as a single-pulse limit for damage to occur which was in the range of 40000 - 45000 *psi* for a 0.5 μm gold layer thickness.

In the case of failure caused by fatigue, the stress level to cause failure for gold was calculated to be 10000 - 12000 *psi*, which translates into a maximum temperature rise of 26-31 K or 5-6 $mJ/cm^2/pulse$ at the resist or an incidence of 14-17 $mJ/cm^2/pulse$ at the mask membrane. Berylium, when considered for membrane material, can reduce shear stress by a factor of 2.1 as compared to silicon. This result is attributed to the fact that berylium has a thermal expansion close to gold and has a lower absorption coefficient than silicon which leads to lower temperatures.

Vladimirsky et al. (1989) considered the thermoelastic plane strain-stress model

$$\frac{\partial^2 \delta}{\partial x^2} = a(1+v)\frac{\partial \Delta T}{\partial x} \tag{2.2.21}$$

where δ is the induced thermoelastic distortion, a is the heat expansion coefficient, v is the Poisson's ratio, and x the position of the beam on the mask membrane (here it is assumed that the x-ray beam is moving in the x direction). Solution of Eq. (2.2.21) for the boundary conditions $x = 0$, $\delta = 0$; $x = L$, $\delta = 0$ gives the induced distortion at point x

$$\delta(x) = a(1+v)\int_0^x \Delta T(\xi - X)d\xi - a(1+v)\frac{x}{L}\int_0^L \Delta T(\xi - X)d\xi. \tag{2.2.22}$$

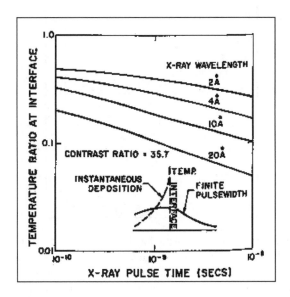

Fig. 2.2.1. Temperature rise at the interface, normalized to the maximum rise, $\Delta T_{\max}$, as a function of pulse duration for different wavelengths (reprinted from Ballantyne and Hyman 1985 with permission from the American Institute of Physics)

If the temperature distribution is constant over the mask ($\Delta T = \Delta T_c$), it is clear from Eq. (2.2.22) that $\delta(x) = a(1+v)[\Delta T_c x - (x/L)\Delta T_c L] = 0$. Figure (2.2.2) presents the mask membrane distortions, $\delta(x)$, as a function of x. For the calculations, the width of the Si mask membrane was 25 mm, $a = 2.6 \times 10^{-6}$ /^0C, $v = 0.25$. The standard deviation for the power was $\sigma_p = 2.5$ mm, and for the temperature distribution was $\sigma_T = 5.0$ mm. It is seen from the figure that, during beam scanning from 0 to 25 mm across the mask, the distribution of the thermoelastic distortions of the membrane is not stable, but changes with the position of the x-ray beam in the exposure field. This movement in distortion is expected to affect the image printed on the resist. The shift or distortion in the printed image $\delta_{im}(x)$ was calculated as the average of $\delta(x)$ with regard to the energy distribution $f(x)$ of the x-ray beam,

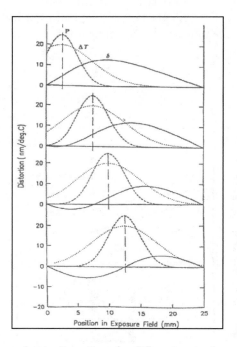

Fig. 2.2.2. Thermoelastic distortions for different x-ray beam positions in the exposure field. The Si mask for this calculation was 25 mm wide and the thermal expansion coefficient $\alpha=2.6\times10^{-6}/^0\mathrm{C}$ with the Poisson's ratio v=0.25. The standard deviation for incident power and temperature distribution were σ_p=2.5mm and σ_T =5.0mm, respectively (reprinted from Vladimirsky et al. 1989 with permission from the American Vacuum Society)

$$\delta_{im}(x) = E(\delta(x) = \int_{-A}^{A} \delta(x)f(x)dx / \int_{-A}^{A} f(x)dx. \qquad (2.2.23)$$

For an exposed area relatively far from the edges of the membrane, an approximate expression for image distortion is given by

$$\delta_{im}(x) \approx a(1+v)\Delta T_{\max}\sigma_T(1-2x/L). \qquad (2.2.24)$$

A maximum displacement in the image ($\delta_{im}^{\max} \approx \alpha(1+v)\Delta T_{\max}\times\sigma_T(1-\frac{3\sigma_T}{L})$) can be expected at $x = 1.5\sigma_T$, and $x = L-1.5\sigma_T$. Additionally, the edge blur can be expressed as $\sigma_{blur}(x) = \sqrt{\langle \delta^2(x)\rangle - \langle \delta(x)\rangle^2}$. A maximum value for the blur is given as

$$\sigma_{blurr}^{\max} = a(1+v)\Delta T \max \frac{\sigma_T \sigma_T}{\sqrt{\sigma_p^2 + \sigma_T^2}}. \tag{2.2.25}$$

Maximum image distortion and image blur were calculated for a 25 mm wide silicon membrane with $a = 2.6x10^{-6}$ $/^0C$, $\sigma_T = 5$ mm , and $\sigma_p = 2.5$ mm. These gave $\delta_{im}^{\max}(x) = 6.5$ $nm/^0C$, and $\sigma_{blurr}^{\max} = 7.2$ $nm/^0C$. Experimental results on thermoelastic distortions using scanning electron microscopy (SEM) showed detectable effects when the mask-wafer assembly was operated in a vacuum. However, for the low-pressure helium atmosphere, distortion effects were negligible.

Computer simulation, using NASTRAN and CAEDS finite element software, was carried out to determine the effect of short pulse irradiation on the thermal and thermoelastic behavior of different mask structures, Si-Au, Si-Polyimide-Au, Si-W, Si-Ta, SIC-Au, SIC-W, SIC-Ta, and Diamond(C)-Au (Shareef et al. 1989, 1990). Two wavelengths, 8.3 and 10 Angstroms, were assumed with a total energy of 10 mJ/cm^2 incident on the mask. The mask was 2 μm thick and the absorber 0.2 μm in thickness. The gap was 40 μm with He at atmospheric pressure. Results indicated that a maximum temperature rise of 31.6 ^0C occurred for the Si-Au x-ray mask structure with 30% absorber coverage at a pulse of 2 ns and at the 8.3 Angstrom wavelength. Maximum rise in temperature did not increase with an increase in membrane coverage. The temperature rise decreased and reached the ambient temperature after about 10 ms which indicated that it was possible to operate at 10 mJ/cm^2 and 100 Hz repetition rate without causing significant heating of the mask. The maximum Von Mises stress was 3.05×10^8 dyn/cm^2, at the 2 ns pulse and 8.3 Angstrom wavelength, for the SIC-Au mask structure. Both temperature rise and stress decreased with an increase in wavelength as well as pulse width. The maximum stress was of the same order as the intrinsic stress level of the mask material. One may assume that, with x-ray pulses higher than 10 mJ/cm^2 in intensity, maximum induced stress will exceed the intrinsic stress, thus leading to distortions in the mask and to mask lifetime problems. Results indicate that it is possible to reduce temperature rise and stress by proper selection of the pulse wavelength and the material for absorber and membrane.

For pulsed x-ray exposure with the interpulse duration much longer than the pulse duration, one may express the power intensity per pulse over time in terms of a Fourier series (Chiba and Okada 1990) which gives

$$I(t)/I_0 = A_0/2 + \sum A_n \cos(n\omega t) + B_n \sin(n\omega t), \tag{2.2.26a}$$

where

$$A_0 = \frac{2}{T_p} \int_0^{T_p} I(t)dt, \tag{2.2.26b}$$

$$A_n = \frac{2}{T_p} \int_0^{T_p} I(t) \cos(n\omega t) dt, \tag{2.2.26c}$$

$$B_n = \frac{2}{T_p} \int_0^{T_p} I(t) \sin(n\omega t) dt, \tag{2.2.26d}$$

$$\omega = 2\pi/T_p.$$

Here, the pulse intensity is approximated by the expression

$$I(t) = I_0 \sin(\pi t/\tau_p) \quad 0 \le t \le \tau_p \tag{2.2.26e}$$
$$= 0 \qquad\qquad\quad \tau_p \le t \le T_p,$$

where

$$I_0 = (\frac{\pi}{2\tau_p})J_0, \tag{2.2.26f}$$

τ_p is the pulse duration, T_p is the period, and J_0 is the energy density per pulse.

The heat source $q(t)$ in the membrane is given, as an average along the membrane depth d, by

$$q(t) = \frac{I(t)}{d} [1 - \exp(-\mu d)], \tag{2.2.27}$$

where μ is the absorption coefficient of the pulse intensity. The heat equation for the temperature distribution is given by

$$\frac{\partial T}{\partial t} = \frac{K}{\rho c}(\frac{\partial^2 T}{\partial x^2} - \gamma T) + \frac{q(t)}{\rho c} - \frac{\epsilon_0 \beta}{\rho c d} [(T_0 + T)^4 - T^4], \tag{2.2.28}$$

where $\gamma = \frac{h_m}{Kd}$, with initial condition $T = 0$ at $t = 0$ and boundary conditions

$$A_1 K_1 \frac{\partial T}{\partial x} = \frac{A_2 K_2}{L}T \quad \text{at} \quad x = 0, \tag{2.2.29}$$

and

$$-A_1 K_1 \frac{\partial T}{\partial x} = \frac{A_2 K_2}{L}T \quad \text{at} \quad x = X_0. \tag{2.2.30}$$

Here, T is the temperature change, T_0 is the initial temperature, K is the thermal conductivity, ρ is the density, c is the specific heat, h_m is the heat transfer coefficient caused by convection, ϵ_0 is the emissivity, β is the Stefan-Boltzmann constant ($5.67 \times 10^{-12} \ W \cdot cm^{-2} K^{-4}$), X_0 is the window size, L is the frame width, and A_1 and A_2 are the membrane and frame thicknesses, respectively.

Distortion as a function of time can be analyzed using the Hamilton's principle of minimizing the time integral of the difference between the kinetic energy T_k and the potential energy U or the elastic strain energy, namely $I = \int (T_k - U)dt$. Here $U = \int 0.5\sigma_x \xi_x dx$, where σ_x, ξ_x are the stress and strain in the x direction, respectively. The strain acting on the membrane in the x direction can be expressed as

$$\xi_x = \frac{\sigma_x - v\sigma_y}{E} + aT(x,t), \qquad (2.2.31)$$

where

$$\sigma_x = \frac{E}{(1-v^2)}[\xi_x - (1+v)aT(x,t)],$$

and

$$\sigma_y = \frac{E}{(1-v^2)}[v\xi_x - (1+v)aT(x,t)]$$

are the stresses in the x and y directions, respectively. As such, U and T_k may be expressed as

$$U = \int 0.5\frac{E}{(1-v^2)}[\xi_x^2 - (1+v)aT(x,t)\xi_x]dx, \qquad (2.2.32)$$

and

$$T_k = \int 0.5\frac{\rho}{g}(\frac{\partial\mu}{\partial t})^2 dx, \qquad (2.2.33)$$

where μ is the displacement or distortion, and g is the acceleration caused by gravity. Expressing the strain as $\xi_x = \frac{\partial\mu}{\partial x} = \mu_x$, one has that

$$I(\mu) = \int (T_k - U)dt \qquad (2.2.34)$$

$$= \int \int F(x,t,\mu,\mu_x,\mu_t)dxdt$$

$$= \int \int [0.5\frac{\rho}{g}(\frac{\partial\mu}{\partial t})^2 - 0.5\frac{E}{(1-v^2)}[\xi_x - (1+v)aT(x,t)\xi_x]dxdt.$$

The solution to the double integral is the function $\mu(x,t)$ that minimizes $I(\mu)$. Setting $\frac{dI(\mu)}{d\mu} = 0$, one obtains the Eular-Lagrange partial differential equation

$$\frac{\partial^2 F}{\partial t\partial\mu_t} = \frac{\partial F}{\partial\mu} - \frac{\partial^2 F}{\partial_x\partial\mu_x} \qquad (2.2.35)$$

From Eq. (2.2.35), one can obtain the thermoelastic wave equation,

$$\frac{\partial^2\mu}{\partial t^2} = C_L^2\left[\frac{\partial^2\mu}{\partial x^2} - (1+v)a\frac{\partial T}{\partial x}\right], \qquad (2.2.36)$$

with initial and boundary conditions,

$$t = 0 \qquad \mu = 0 \qquad \frac{\partial\mu}{\partial t} = 0,$$

$$x = 0 \qquad x = X_0 \qquad \mu = 0.$$

Here, $C_L^2 = gEv/\rho$, where C_L is the longitudinal wave speed in materials. When the propagation time across the membrane is much shorter than the pulse duration, thermal distortion can be treated by the classical static thermoelastic methodology. However, in the case of pulsed x-ray with very short pulse duration (shorter than the wave propagation time through the material), one expects an effect of the elastic wave on thermal distortion.

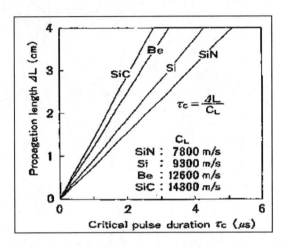

Fig. 2.2.3. Relationship between propagation length and critical pulse duration in the case of SiN, Si, Be, and SiC (reprinted from Chiba and Okada 1990 with permission from the Japanese Institute of Pure and Applied Physics)

Equations (2.2.28) and (2.2.36) were solved numerically using the Crank-Nicolson method. This solution was performed in the case of pulsed x-ray for SiN, Si, SiC, and Be membranes of 2 μm thickness and 4 cm window size under and a 1 bar He gas. The incident pulse intensity was considered to be 30 mJ/cm^2. In view of the fact that the effect of the elastic wave on thermal distortion depends on the wave propagation time relative to the pulse duration time, one may define a critical pulse duration (CPD) by $\tau_c = \Delta L/C_L$, where ΔL is a distance from the rim of the membrane toward the center of the membrane. It should be noted that the wave propagates from the rim toward the center of the membrane. If the pulse duration is shorter than the CPD, the elastic wave falls short of reaching a point x a distance ΔL from the rim during the pulse exposure. As a result, one does not expect any dynamic thermal distortion based on the propagation of the elastic wave at point x during the pulse exposure. However, static or quasi-static distortion may occur. Since C_L decreases with a decrease in Young's modulus E, one expects a mask membrane with a low Young's modulus (or a low C_L and therefore a high CPD) to be more suitable for pulsed x-ray exposures. Figure (2.2.3) shows the relationship between critical pulse duration and propagation length for different membrane materials. For a given propagation length ΔL, SiN has longer CPD time than Si, Be, or SiC. Be and SiC are not adequate because of short CPD time. Hence, for pulsed x-ray exposure, critical pulse duration seems to be an important factor in choosing the mask membrane. Figure (2.2.4) presents thermal peak displacement and time lag for different pulse durations. It is seen that the peak displacement (or distortion) increases with a decrease in pulse duration and reaches a constant value of about 33 nm when the pulse duration is under 100 ms. On the other hand, the time lag decreases with a decrease in pulse duration time and converges to a value of about 23 nm under 10 ms pulse duration. For pulse duration above 100 ms, it is seen that the time lag is approximately equal to the pulse duration implying that the peak displacement occurs at the end of the pulse duration. This result is true also for peak temperature as seen in Fig. (2.2.5). It is seen from this figure that, for pulses under 10 ms in duration, peak thermal stress values converge to -133kg/cm^2 and peak temperature values converge to 24 degrees.

Using a similar approach as for pulsed x-ray exposure, Chiba and Okada (1991) considered thermal distortion of a mask for synchrotron radiation. Their analysis considered temperature distributions in the membrane, in the He gas of the proximity gap, and in the resist (Fig. 2.2.6). The He gas under 760 Torr pressure was considered to be characterized by viscous flow. This is a true assessment for gaps larger than 10 μm based on the Knudsen number. The heat balance equations, taking into effect heat sources, radiation, convection, conduction, and heat transfer, are given by

$$\frac{\partial T_1}{\partial t} = k_1 \frac{\partial^2 T_1}{\partial x^2} + \frac{Q_1(x,t)}{\rho_1 c_1} - \frac{(h + a_r)}{\rho_1 c_1 d_1} T_1 - \frac{h_{1,2}}{\rho_1 c_1 d_1}(T_1 - T_2), \qquad (2.2.37a)$$

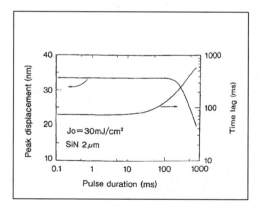

Fig. 2.2.4. Effect of pulse duration on peak displacement, and time lag (reprinted from Chiba and Okada 1990 with permission from the Japanese Institute of Pure and Applied Physics)

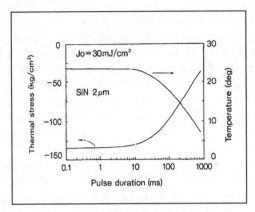

Fig. 2.2.5. Effect of pulse duration on peak temperature and peak thermal stress (reprinted from Chiba and Okada 1990 with permission from the Japanese Institute of Pure and Applied physics)

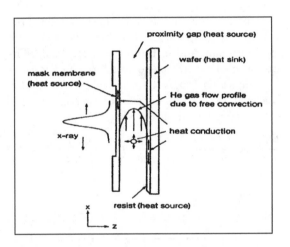

Fig. 2.2.6. Schematic diagram of a synchrotron radiation lithography set up used in the thermal distortion model (reprinted from Chiba and Okada 1991 with permission from the American Vacuum Society)

$$\frac{\partial T_2}{\partial t} + U\frac{\partial T_2}{\partial x} = k_2\frac{\partial^2 T_2}{\partial x^2} + \frac{Q_2(x,t)}{\rho_2 c_2}$$
$$+ \frac{h_{1,2}}{\rho_2 c_2 d_2}(T_1 - T_2) + \frac{h_{2,3}}{\rho_2 c_2 d_2}(T_3 - T_2), \quad (2.2.37\mathrm{b})$$

$$\frac{\partial T_3}{\partial t} = k_3 \frac{\partial^2 T_3}{\partial x^2} + \frac{Q_3(x,t)}{\rho_3 c_3} - \frac{h_{2,3}}{\rho_3 c_3 d_3}(T_3 - T_2) - \frac{h_{3,4}}{\rho_3 c_3 d_3}T_3, \qquad (2.2.37c)$$

where

$$a_r = 4\varepsilon_0 \beta T_0^3, \qquad h_{1,2} = hc + \frac{K_2}{d_2} = h_{2,3},$$

$$hc = \frac{g\beta c T_2 K_2 \Pr d_2^3}{4X_0 v_f}, \qquad \beta c = \frac{1}{(273 + T_2)},$$

$k_i = \frac{K_i}{\rho_i c_i}$ ($i = 1, 2, 3$) is the thermal diffusivity, K_i is thermal conductivity, ρ_i is density, c_i is specific heat, β is the Stefan-Boltzmann constant , g is the acceleration due to gravity, βc is the compressibility, v_f is the kinetic viscosity, a_r is the radiation heat transfer coefficient, $\Pr$ is the Prandtl number, U is the free convection velocity in the vertical direction, h is the heat transfer coefficient describing heat flow from the membrane to the surrounding gas on the opposite side of the gap, Q_i is the heat source from x-ray exposure, $h_{1,2}$ and $h_{2,3}$, represent heat transfer and heat conduction from region 1 to 2 and from 2 to 3, and $h_{3,4}$ is the heat transfer from the resist to the wafer or substrate. The heat sources are determined by the energy absorbed by each of the membrane, gas in the gap, and resist. These sources are represented by

$$Q_1(x,t) = \frac{I_0}{d_1}[1 - \exp(-\mu_1 d_1)]f(x,t), \qquad (2.2.38)$$

$$Q_2(x,t) = \frac{I_0 \exp(-\mu_1 d_1)}{d_2}[1 - \exp(-\mu_2 d_2)]f(x,t), \qquad (2.2.39)$$

$$Q_3(x,t) = \frac{I_0 \exp(-\mu_1 d_1 - \mu_2 d_2)}{d_3}[1 - \exp(-\mu_3 d_3)]f(x,t), \qquad (2.2.40)$$

where

$$f(x,t) = \exp[-\frac{(x - x_s - V_0 t)^2}{2R_0^2}], \qquad (2.2.41)$$

x_s is the exposure starting point on the mask membrane, V_0 is the scan speed, and R_0 is the beam radius on the membrane. The equation for the convection velocity U is given by

$$\frac{\partial U}{\partial t} + \frac{U}{5}\frac{\partial U}{\partial x} = v_f(\frac{\partial^2 U}{\partial x^2} - \frac{12}{d_2^2}U) + 6g\beta c T_2. \qquad (2.2.42)$$

In order to determine the temperature distribution in the membrane, one must solve Eqs. (2.2.37) and (2.2.42) simultaneously. The elastic wave equation for displacement with an added frictional factor γ whose effect is to diminish the movement of the membrane caused by transient thermal expansion can be obtained from Eq. (2.2.36). This gives

$$\frac{\partial^2 \mu}{\partial t^2} + 2\gamma \frac{\partial \mu}{\partial t} = C_L^2 \left[\frac{\partial^2 \mu}{\partial x^2} - (1 + v)a\frac{\partial T_1}{\partial x}\right], \qquad (2.2.43)$$

where $C_L = \sqrt{\frac{gE}{\rho(1-v^2)}}$ is the longitudinal propagation velocity of the wave through the membrane.

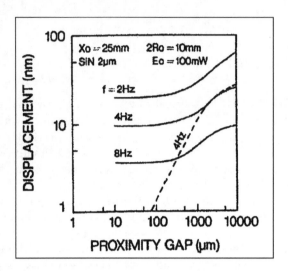

Fig. 2.2.7. Effect of proximity gap on displacement at the center of the membrane for different scanning exposure frequencies under a 760-Torr helium gas environment. The dotted line is based on the conventional model. The membrane window size is denoted by X_0, E_0 is the beam power, and $2R_0$ is the x-ray beam size (reprinted from Chiba and Okada 1991 with permission from the American Vacuum Society)

The effect of the proximity gap on the displacement at the center of the membrane is shown in Fig. (2.2.7). The dotted line is for comparison and represents results based on the conventional model where the heat equation was considered only in the membrane (Heinrich et al., 1983 ; Vladimirsky et al., 1989). It is seen that for the present model, unlike the conventional model, displacement converged to a constant value, depending on the scan frequency, in the gap range 10 to 100 μm. Figure (2.2.8) gives the effect of scan frequency on the behavior of the maximum temperature in the whole membrane and maximum displacement in the whole membrane and at the center (U_c). It is seen that above a 1 Hz frequency both displacement and temperature decreased with an increase in frequency. Above a 10 Hz frequency, Umax and U_c diverged considerably. The effect of window size on maximum displacement (distortion) for several constant scan speeds is shown in Figure (2.2.9). It is seen from this figure that displacement increased with a decrease in scan speed. For a scan speed of 200 cm/s, displacement increased with an increase in window size, then decreased slightly and eventually converged to a constant value. For scan speeds of 2 cm/s and 10 cm/s, displacement increased with an increase in window size. However, displacement is expected to converge to a constant value that depends on the scan speed.

The temperature rise in an x-ray mask has been modeled by Li et al. (1995) using the heat Eq. (2.2.28). The heat source in the equation for the membrane, absorber, and proximity gap was given as

$$H_1 = \frac{I_0}{d_1}[1 - \exp(-\mu_1 d_1)], \tag{2.2.44}$$

$$H_2 = \frac{I_0 \exp(-\mu_1 d_1)}{d_2}[1 - \exp(-\mu_2 d_2)], \tag{2.2.45}$$

and

$$H_3 = \frac{I_0 \exp(-\mu_1 d_1 - \mu_2 d_2)}{d_3}[1 - \exp(-\mu_3 d_3)], \tag{2.2.46}$$

respectively.

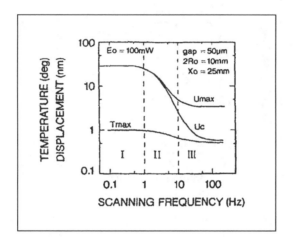

Fig. 2.2.8. Effect of scan frequency on maximum temperature rise and maximum displacement. U_C is displacement at the membrane center (reprinted from Chiba and Okada 1991 with permission from the American Vacuum Society)

Because the membrane's thickness is small relative to the mask window size, the mask membrane was assumed to be only in a plane stress. Equations relating stresses to the strains, $\xi_x, \xi_y, \gamma_{xy}$, are

$$\xi_x = \frac{\partial u}{\partial x} = \frac{1}{E}(\sigma_x - v\sigma_y) + aT(x, y, t), \tag{2.2.47}$$

$$\xi_y = \frac{\partial v}{\partial y} = \frac{1}{E}(\sigma_y - v\sigma_x) + aT(x, y, t), \tag{2.2.48}$$

$$\gamma_{xy} = \frac{\partial u}{\partial x} + \frac{\partial v}{\partial y} = \frac{2(1+v)}{E}\tau_{xy}. \qquad (2.2.49)$$

Using a finite element program (MSC/NASTRAN), temperature distributions, thermal stresses, and strains and thermal displacements in a silicontungsten (Si-W) mask structure were presented. The Si membrane considered was rectangular, 6×7 μm with 2 μm thickness, and the W absorber was 0.8 μm thick with a minimum width of 0.25 μm. Results indicated that maximium temperature rises occurred, as expected, in regions of dense absorber patterns. Also, thermal stresses were predominant at the Si-W interface and at the edges of the absorber. Considering thermal displacements, it was concluded that the temperature rise should not exceed 3.42 ^0C in order for the distortion to remain within the acceptable limit of 25 nm.

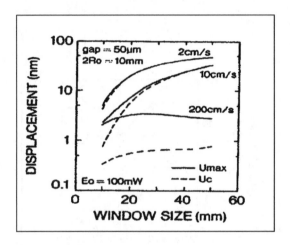

Fig. 2.2.9. Effect of membrane window size on displacement for different scan speeds (reprinted from Chiba and Okada 1991 with permission from the American Vacuum Society)

Temperature rise in the mask and thermal distortions in the absorber and membrane have been modeled (Lakhsasi et al. 1997) using the three-dimensional heat equation and thermal stresses and strains in a plane. Model calculations were obtained by means of finite element analysis using the commercially available computer code DISPGDP. The model was applied to a mask structure, under a helium gas environment, with silicon carbide (SiC) as the membrane and with tungsten (W) as the absorber. The heat source was due to short x-ray pulses from a laser plasma. The effects of SiC thermal conductivity, pulse length, x-ray wavelength, and pulse repetition rate on the thermal behavior of the membrane and absorber were presented. It was shown that, for a 2 ns pulse duration with a laser intensity of 10 mJ/cm^2 and a wavelength of 8.3 Angstrom, the maximum temperature rise occurred at the end duration of the x-ray pulse and was highest at the SiC-W interface for low SiC conductivity of 0.41 W/cm^0C. At 1.9 and 3.5 W/cm^0C SiC conductivity, the maximum temperature rise shifted to the W-He interface. After the maximum rise, the temperature decreased and reached the ambient temperature in 5 ms for the low thermal conductivity and in 10 ms for the higher conductivities. Hence, the cooling time can be used to determine the pulse repetition rate in order not to cause any significant heating in the mask. The rates in this case can be 200 Hz or 100 Hz for the 5 ms and 10 ms, respectively. There were no substantial effects of pulse length, amplitude, or wavelength on thermal stress and distortion. On the other hand, it was determined that percent deformation in the absorber increased significantly with an increase in the SiC thermal conductivity. This is explained by the fact that for high SiC thermal conductivity heat accumulates in the absorber. Therefore, a very low membrane thermal conductivity is expected to minimize thermal stress and deformation in the absorber.

2.3 Thermal Analysis of Resist

Polymethylmethacrylate (PMMA) and other photoresists used in x-ray lithography for the manufacturing of high aspect ratio microstructures have low heat conductivities by comparison to the supporting silicon substrate. For instance, the thermal conductivity of PMMA is $k = 0.198 \ W/m/K$ while the silicon substrate has a conductivity of $k = 148 \ W/m/K$. This condition can cause heating and thermal expansion in the resist when it is exposed to high flux synchrotron radiation which may lead to distortions in the exposed area. Additionally, heat expansion can lead to problems at the resist-silicon interface where bonding is weak. One- and two-dimensional heat conduction equations were used by Ameel et al. (1994) to study temperature rise and thermal displacement in a PMMA resist on a silicon substrate. The analysis considered a circular resist and substrate of radius $a = 1 \ mm$ and a circular mask induced pattern on the resist of radius $a = 5 \ mm$. Each resist and substrate had a thickness of 100 μm. Figure (2.3.1) gives a schematic diagram of the resist and substrate configuration. The resist is assumed to be uniformly irradiated with an intensity I_0. The heat source at depth z is expressed as $g = I_0\mu \exp(-\mu z)$ where μ is the linear absorption coefficient in the resist. The substrate is assumed to be cooled and remains at a fixed temperature taken to be zero. The one-dimensional heat equation in the z-direction with a heat source (g) was solved at steady state for three boundary conditions at $z = 0$ (exposed surface of the resist). The solutions are given by

$$T(z) = \frac{I_0}{k}\left(\frac{e^{-\mu L}}{\mu} + L - \frac{e^{-\mu z}}{\mu} - z\right); \qquad B.C: -k\frac{dT}{dz} = 0, \quad \text{adiabatic} \quad (2.3.1)$$

$$T(z) = \frac{I_0}{k\mu}\left(1 - e^{-\mu z} - \frac{z}{L}(1 - e^{-L})\right); \qquad B.C: T = 0, \quad \text{homogeneous} \quad (2.3.2)$$

$$T(z) = \frac{-I_0}{k\mu}e^{-\mu z} + \frac{\frac{-h_c I_0}{k\mu}(e^{-\mu L} - 1) - I_0}{h_c L + k}z + \frac{I_0}{k\mu}e^{-\mu L}$$
$$-\frac{\frac{-h_c I_0}{k\mu}(e^{-\mu L} - 1) - I_0}{h_c L + k}L;$$

$$B.C : k\frac{dT}{dz} = h_c T, \quad \text{forced convection.} \tag{2.3.3}$$

Here, $h_c = 60 \ W/cm^2/K$ = convective heat transfer coefficient.

Based on these equations, the maximum temperature rise was 11.77 K, 2.26K, and 11.27K for the adiabatic, homogeneous, and forced convection boundary conditions, respectively. Assuming radial symmetry, a 2D heat equation in the z- and radial directions was solved at steady state using the finite element software COSMOS. The temperature profile at the center line in the z-direction is presented in Fig. (2.3.2) As expected, temperature rise is maximum at the surface and decreases to zero, the temperature of the substrate, at the interface. A solution in the unsteady state revealed that the temperature reached 95% of its steady state value in approximately 0.125 s. Based on the temperature rise, the maximum distortion in the resist was approximately 0.1 μm. This result was considered to be a limiting factor in the production of deep aspect ratio (over 100 μm) microstructures with high accuracy or small tolerance.

The above 2D heat conduction model can be extended to three dimensions. Figure 2.3.3 gives the 3-D configuration of resist and substrate. In the 3-D case, the heat equations for resist and substrate are expressed as

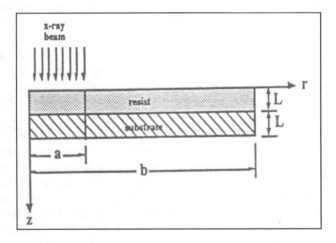

Fig. 2.3.1. Schematic diagram of resist and substrate used in the analysis (reprinted from Ameel et al. 1994 by permission of Taylor & Francis Ltd. http://www.tandf.co.uk/journals)

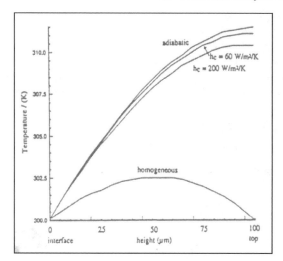

Fig. 2.3.2. Profiles of temperature rise at the center line for different boundary conditions (reprinted from Ameel et al. 1994 by permission of Taylor & Francis Ltd. http://www.tandf.co.uk/journals)

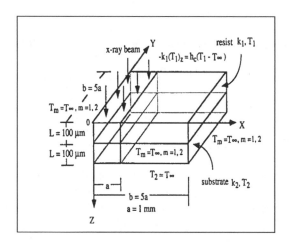

Fig. 2.3.3. Schemetic diagram of a resist and substrate configuration (Reproduced from Dai et al. 1997a by permission of Taylor & Francis Inc., http://www.routledge-ny.com)

$$\rho_1 c_1 \frac{\partial T_1}{\partial t} = k_1 \left(\frac{\partial^2 T_1}{\partial x^2} + \frac{\partial^2 T_1}{\partial y^2} + \frac{\partial^2 T_1}{\partial z^2} \right) + g(x, y, z, t) \qquad (2.3.4)$$

and

$$\rho_2 c_2 \frac{\partial T_2}{\partial t} = k_2 \left(\frac{\partial^2 T_2}{\partial x^2} + \frac{\partial^2 T_2}{\partial y^2} + \frac{\partial^2 T_2}{\partial z^2} \right) + g(x, y, z, t) \qquad (2.3.5)$$

where $g(x, y, z, t)$ is a heat source term and $T_1, T_2, k_1, k_2, \rho_1, \rho_2$, and c_1, c_2 are temperature rise, thermal conductivities, densities, and specific heat for the resist and substrate, respectively (Dai et al. 1997a). The source term is given by

$$g = I_0 \mu \exp(-\mu z); \quad 0 \le x \le a \quad 0 \le y \le b$$
$$g = 0 \qquad a < x \le b \quad 0 \le y \le b \qquad (2.3.6)$$

The boundary condition at the interface between resist and substrate is

$$-k_1 \frac{\partial T_1}{\partial z} = -k_2 \frac{\partial T_2}{\partial z} \quad ; \quad T_1 = T_2$$

and

$$T_i = T_\infty, \quad i = 1, 2$$

at the side walls of the resist and substrate and $T_2 = T_\infty$ at the bottom surface of the substrate, where T_∞ is the surrounding temperature. On the surface of the resist exposed to helium in the proximity gap, the boundary condition is

$$k_1 \frac{\partial T_1}{\partial z} = h_c (T_1 - T_\infty). \qquad (2.3.7)$$

Let T_{ijk}^n denote the numerical approximation to the exact solution $T(i\Delta x, j\Delta y, k\Delta z, n\Delta t)$ where Δt is the time increment and $\Delta x, \Delta y$, and Δz are the grid sizes in the x, y, and z directions, respectively. The same mesh points are chosen for each layer. Also, the centered-difference formula $\frac{1}{\Delta x^2} \delta_x^2 T_{ijk}^n$ was used to approximate $\frac{\partial^2 T}{\partial x^2}$, where

$$\delta_x^2 T_{ijk}^n = T_{i+1jk}^n - 2T_{ijk}^n + T_{i-1jk}^n$$

and similarly for $\frac{\partial^2 T}{\partial y^2}$ and $\frac{\partial^2 T}{\partial z^2}$. As such, the numerical model to solve the heat equations can be developed by applying the three-dimensional Douglas ADI method. This process leads to the following computational procedure. First, obtain the temperatures $(T_1)_{ijk}^{n+1/3}$ and $(T_2)_{ijk}^{n+1/3}$ from the two tridiagonal linear systems

$$-b_k^1 (T_1)_{ijk-1}^{n+1/3} + a_k^1 (T_1)_{ijk}^{n+1/3} - c_k^1 (T_1)_{ijk+1}^{n+1/3} = d_{ijk}^1 \qquad (2.3.8a)$$

$$-b_k^2 (T_2)_{ijk-1}^{n+1/3} + a_k^2 (T_2)_{ijk}^{n+1/3} - c_k^2 (T_2)_{ijk+1}^{n+1/3} = d_{ijk}^2 \qquad (2.3.8b)$$

where $i = 1, 2, ..., N_x$; $j = 1, 2, ..., N_y$; and $k = 1, 2, ..., N_z$. Because the temperature at the $(n+1/3)\Delta t$ time step is unknown at the interface between resist and substrate, Eqs. (2.3.8a) and (2.3.8b) can be solved by using the so-called "divide and conquer" technique which is accomplished by using the two iterative formulae

$$(T_1)_{ijk}^{n+1/3} = \beta_k^1 (T_1)_{ijk+1}^{n+1/3} + v_k^1 \qquad k = N_z, ..., 1 \tag{2.3.9a}$$

$$(T_2)_{ijk}^{n+1/3} = \beta_k^2 (T_2)_{ijk-1}^{n+1/3} + v_k^2 \qquad k = 1, ..., N_z \tag{2.3.9b}$$

where

$$\beta_k^1 = \frac{c_k^1}{a_k^1 - b_k^1 \beta_{k-1}^1} \qquad \beta_0^1 = 0,$$

$$v_k^1 = \frac{d_{ijk}^1 + b_k^1 v_{k-1}^1}{a_k^1 - b_k^1 \beta_{k-1}^1} \qquad v_0^1 = 0 \qquad k = 1, ..., N_z,$$

$$\beta_k^2 = \frac{b_k^2}{a_k^2 - c_k^2 \beta_{k+1}^2} \qquad \beta_{N_z+1}^2 = 0,$$

and

$$v_k^2 = \frac{d_{ijk}^2 + c_k^2 v_{k+1}^2}{a_k^2 - c_k^2 \beta_{k+1}^2} \qquad v_{N_z+1}^2 = 0 \qquad k = N_z, ..., 1.$$

Then,

$$(T_1)_{ijN_z}^{n+1/3} = \beta_{N_z}^1 (T_1)_{ijN_z+1}^{n+1/3} + v_{N_z}^1, \tag{2.3.10a}$$

and

$$(T_2)_{ij1}^{n+1/3} = \beta_1^2 (T_2)_{ij0}^{n+1/3} + v_1^2 \tag{2.3.10b}$$

are substituted into the interfacial boundary equation

$$-k_1 \left(\frac{(T_1)_{ijN_z+1}^n - (T_1)_{ijN_z}^n}{\Delta z} \right) = -k_2 \left(\frac{(T_2)_{ij1}^n - (T_2)_{ij0}^n}{\Delta z} \right) \tag{2.3.11}$$

to solve for $(T_1)_{ijN_z+1}^{n+1/3}$ and $(T_2)_{ij0}^{n+1/3}$. Once these are obtained, the temperatures $(T_1)_{ijk}^{n+1/3}$ and $(T_2)_{ijk}^{n+1/3}$ are obtained from Eqs. (2.3.9a) and (2.3.9b). The second step in the solution is to solve for the temperatures $(T_1)_{ijk}^{n+2/3}$ and $(T_2)_{ijk}^{n+2/3}$ by using the following equations:

$$(T_1)_{ijk}^{n+2/3} - (T_1)_{ijk}^{n+1/3} = \frac{r_x^1}{2} \delta_x^2 \left[(T_1)_{ijk}^{n+2/3} - (T_1)_{ijk}^n \right], \tag{2.3.12a}$$

$$(T_2)_{ijk}^{n+2/3} - (T_2)_{ijk}^{n+1/3} = \frac{r_x^2}{2} \delta_x^2 \left[(T_2)_{ijk}^{n+2/3} - (T_2)_{ijk}^n \right], \tag{2.3.12b}$$

where $r_x^1 = \frac{k_1 \Delta t}{\rho_1 c_1 \Delta x^2}$ and $r_x^2 = \frac{k_2 \Delta t}{\rho_2 c_2 \Delta x^2}$. Substituting the values $(T_1)_{ijN_z}^{n+2/3}$ and $(T_2)_{ij1}^{n+2/3}$ into the interfacial boundary Eq. (2.3.11), one obtains $(T_1)_{ijN_z+1}^{n+2/3}$ and $(T_2)_{ij0}^{n+2/3}$. The third step in the computation is to obtain $(T_1)_{ijk}^{n+1}$ and $(T_2)_{ijk}^{n+1}$ by using the equations given by

$$(T_1)_{ijk}^{n+1} - (T_1)_{ijk}^{n+2/3} = \frac{r_y^1}{2} \delta_y^2 \left[(T_1)_{ijk}^{n+1} - (T_1)_{ijk}^n \right], \tag{2.3.13a}$$

$$(T_2)_{ijk}^{n+1} - (T_2)_{ijk}^{n+2/3} = \frac{r_y^2}{2}\delta_y^2 \left[(T_2)_{ijk}^{n+1} - (T_2)_{ijk}^n\right], \qquad (2.3.13b)$$

in a similar way to the second step. In the above equations, $r_y^1 = \frac{k_1 \Delta t}{\rho_1 c_1 \Delta y^2}$ and $r_y^2 = \frac{k_2 \Delta t}{\rho_2 c_2 \Delta y^2}$.

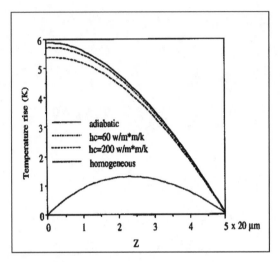

Fig. 2.3.4. Temperature profiles in the resist at x = 0.7 mm and y =2.5 mm for different boundary consditions (Reproduced from Dai et al. 1997a by permission of Taylor & Francis, Inc., http://www.routledge-ny.com)

Numerical results of maximum temperature rise in the resist are presented in Fig. (2.3.4) for the same boundary conditions in Ameel et al. (1994) at $z = 0$. The thermophysical properties (ρ in g/cm^3, k in $W/cm/K$, c_p in $kJ/kg/K$) were 1.18, 0.00198, 1.42 for PMMA and 2.3, 1.5, 0.715 for the silicon substrate. The coefficients I_o and μ in the source term were $3.4W/cm^2$ and $1/106 \times 10^4/cm$, respectively. The convective heat transfer coefficient was $h_c = 60 \ W/m^2/K$, each layer thickness was 100 μm, and the surrounding temperature was $300K$. Under these experimental conditions, it is seen that the maximum temperature rise was $5.72K$ at $x = 0.1$ mm for the adiabatic condition.

The present model can be extended to deal with a circular mask pattern on the resist, as is seen in Fig. (2.3.5). The numerical method for solving Eqs. (2.3.4) and (2.3.5) is based on a hybrid finite element and the generalized Marchuk splitting scheme (Dai and Nassar, 1997b). A domain decomposition algorithm is used with a parallel Gaussian elimination technique to obtain the solution. The finite element is applied first to the circular pattern in the xy cross section. Figure (2.3.6) presents the finite element mesh in the first quadrant. Consider the heat equation

$$\frac{\partial T}{\partial t} = \beta \left(\frac{\partial^2 T}{\partial x^2} + \frac{\partial^2 T}{\partial y^2} + \frac{\partial^2 T}{\partial z^2} \right) + g \qquad (2.3.14)$$

with $T(x, y, z) = 0$ at the side wall. Let

$$\int \int \left[\frac{\partial T}{\partial t} v - \beta \left(\frac{\partial^2 T}{\partial x^2} + \frac{\partial^2 T}{\partial y^2} + \frac{\partial^2 T}{\partial z^2} \right) v - gv \right] dxdy$$

$$= \int \int \left[\frac{\partial T}{\partial t} v + \beta \left(\frac{\partial T}{\partial x} \frac{\partial v}{\partial x} + \frac{\partial T}{\partial y} \frac{\partial v}{\partial y} \right) - \beta \frac{\partial^2 T}{\partial z^2} v - gv \right] dxdy$$

$$= 0 \qquad (2.3.15)$$

where $v(x, y)$ is a function on the Sobolov space H_0^1 (Chandrupatla and Belegundu, 1991). On each element, as shown in Fig. (2.3.7), choose a linear basis function

$$\varphi_p(x, y) = \frac{1}{2S_\Delta} (a_p + b_p x + c_p y) \qquad (2.3.16)$$

where $a_p = x_q y_r - x_r y_q$, $b_p = y_q - y_r$, $c_p = x_r - x_q$, and $2S_\Delta = \frac{1}{2} [x_p (y_q - y_r) + x_q (y_r - y_p) + x_r (y_p - y)]$. Let

$$T(x, y, z, t) = \sum_{p=1}^{N} T_p(z, t) \varphi_p(x, y)$$

and

$$g_h = \sum_{p=1}^{N} g_p \varphi_p(x, y)$$

where N is the number of grid points in the circular xy cross-section. Substituting $T(x, y, z, t)$ and g_h into Eq. (2.3.15) and letting $v = \varphi_q$, one obtains

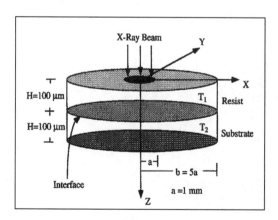

Fig. 2.3.5. Three-dimensional circular configuration of resist and substrate (Reproduced from Dai and Nassar 1997b by permission of Taylor & Francis Inc., http://www.routledge-ny.com)

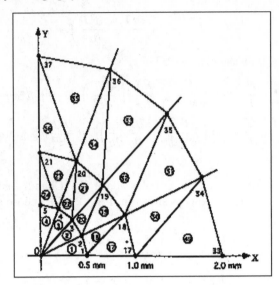

Fig. 2.3.6. Finite element mesh in the first quadrant of the xy cross section (Reproduced from Dai and Nassar 1997b by permission of Taylor & Francis Inc., http://www.routledge-ny.com)

$$\sum_{p=1}^{N} \frac{\partial T_p}{\partial t} \int\int \varphi_p \varphi_q dx dy$$

$$+ \sum_{p=1}^{N} [\beta T_p \int\int \left(\frac{\partial \varphi_p}{\partial x} \frac{\partial \varphi_q}{\partial x} + \frac{\partial \varphi_p}{\partial y} \frac{\partial \varphi_q}{\partial y} \right) dx dy$$

$$- \beta \frac{\partial^2 T_p}{\partial z^2} \int\int \varphi_p \varphi_q dx dy]$$

$$- \sum_{p=1}^{N} g_p \int\int \varphi_p \varphi_q dx dy = 0 ; \qquad q = 1, 2, ..., N \qquad (2.3.17)$$

In matrix notation, Eq. (2.3.17) may be expressed as

$$M \frac{\partial \mathbf{T}}{\partial t} + K\mathbf{T} - \beta M \frac{\partial^2 \mathbf{T}}{\partial z^2} = M\mathbf{f}, \qquad (2.3.18)$$

where $\mathbf{T}(z, t) = [T_1(z, t), ..., T_N(z, t)]^T$, $\mathbf{f}(z, t) = (g_1, ..., g_N)^T$ and the matrices $M_{N \times N}$ and $K_{N \times N}$ have the respective entries

$$m_{qp} = \int\int \varphi_p \varphi_q dx dy \text{ and } k_{qp} = \beta \int\int \left(\frac{\partial \varphi_p}{\partial x} \frac{\partial \varphi_q}{\partial x} + \frac{\partial \varphi_p}{\partial y} \frac{\partial \varphi_q}{\partial y} \right) dx dy.$$

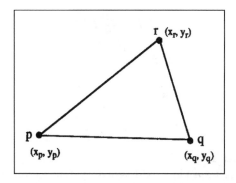

Fig. 2.3.7. Triangular element Δ (Reproduced from Dai and Nassar 1997b by permission of Taylor & Francis Inc., http://www.routledge-ny.com)

Applying the lumped mass technique (Chandrupatla and Belegundu 1991) to obtain a diagonal matrix D and replacing M by D, Eq.(2.3.18) becomes

$$D\frac{\partial \mathbf{T}}{\partial t} + K\mathbf{T} - \beta D \frac{\partial^2 \mathbf{T}}{\partial z^2} = D\mathbf{f}, \qquad (2.3.19)$$

where each entry d_i on the diagonal of D is $\frac{1}{3}\sum_\Delta S_\Delta$ (i.e., one-third of the sum of all elements with node i as one vertex).

Let $\mathbf{T}_m^n = \mathbf{T}(m\Delta z, n\Delta t)$, where Δz is the grid size in the z direction and Δt is the time increment $m = 1, ..., N_t$ and $n\Delta t \leq t_0$. A generalized Marchuk splitting scheme for Eq. (2.3.19) can be written as follows:

$$(D + \frac{\Delta t}{2}K)\mathbf{T}_m^{n-(1/2)} = (D - \frac{\Delta t}{2}K)\mathbf{T}_m^{n-1} \qquad (2.3.20a)$$

$$(D - \frac{\beta \Delta t}{2}D\delta_z^2)(\mathbf{T}_m^n - \Delta t\mathbf{f}_m^n) = (D + \frac{\beta \Delta t}{2}D\delta_z^2)\mathbf{T}_m^{n-(1/2)} \qquad (2.3.20b)$$

$$(D - \frac{\beta \Delta t}{2}D\delta_z^2)\mathbf{T}_m^{n+(1/2)} = (D + \frac{\beta \Delta t}{2}D\delta_z^2)(\mathbf{T}_m^n + \Delta t\mathbf{f}_m^n) \qquad (2.3.20c)$$

$$(D + \frac{\Delta t}{2}K)\mathbf{T}_m^{n+1} = (D - \frac{\Delta t}{2}K)\mathbf{T}_m^{n+(1/2)} \qquad (2.3.20d)$$

where

$$\delta_z^2\mathbf{T} = \frac{1}{\Delta z^2}(\mathbf{T}_{m+1} - 2\mathbf{T}_m + \mathbf{T}_{m-1}); \quad m = 1, ..., N_z$$

The above scheme is second order accurate with respect to Δt and is unconditionally stable with regard to the initial condition and the source term when $0 < t = n\Delta t \leq t_0$.

To solve for $\mathbf{T}_m^{n-(1/2)}$ and $\mathbf{T}_m^{n+1}$ in Eq.(2.3.20a,d), apply the *LT* factorization and let $D + \frac{\Delta t}{2}K = LU$. Then, one obtains

$$LT_m^* = (D - \frac{\Delta t}{2}K)T_m^{n-1}$$

$$UT_m^{n-(1/2)} = T_m^* \qquad (2.3.21a)$$

and

$$LT_m^{**} = (D - \frac{\Delta t}{2}K)T_m^{n+(1/2)}$$

$$UT_m^{n+1} = T_m^{**}. \qquad (2.3.21b)$$

To solve for T_m^n and $T_m^{n+(1/2)}$, multiply Eqs. (2.3.20b,c) by D^{-1} and reduce the equations to solving a block tridiagonal system as follows:

$$-\lambda T_{m-1} + (1 + 2\lambda)T_m - \lambda T_{m+1} = d_m$$

$$T_0 = T_{N_z+1} = 0; \quad m = 1, ..., N_z \qquad (2.3.22)$$

where $\lambda = \beta \Delta t / \Delta z^2$. A computational procedure of the Gaussian elimination technique for solving Eq. (2.3.22) may be formulated as

$$a_0 = 0 \quad v_o = 0$$

$$a_k = [(1 + 2\lambda) - \lambda a_{k-1}]^{-1} \lambda$$

$$v_k = [(1 + 2\lambda) - \lambda a_{k-1}]^{-1} [d_k + \lambda v_{k-1}] \quad k = 1, ..., N_z$$

$$T_m = a_m T_{m+1} + v_{m+1} \quad T_{N_z+1} = 0 \quad m = N_z, ..., 1 \qquad (2.3.23)$$

or

$$\tilde{a}_{N_z+1} = 0 \quad \tilde{v}_0 = 0$$

$$\tilde{a}_k = [(1 + 2\lambda) - \lambda \tilde{a}_{k+1}]^{-1} \lambda$$

$$\tilde{v}_k = [(1 + 2\lambda) + \lambda \tilde{a}_{k+1}]^{-1} [d_k + \lambda \tilde{v}_{k+1}] \quad k = N_z, ..., 1$$

$$\tilde{T}_m = \tilde{a}_m T_{m-1} + \tilde{v}_{m-1} \quad T_0 = 0 \quad m = 1, ..., N_z. \qquad (2.3.24)$$

From Eqs. (2.3.20a,b,c,d), (2.3.21a,b), (2.3.22). (2.3.23), and (2.3.24) one obtains a solution for the temperature rise in the resist and substrate. The profile of the maximum temperature rise in the resist agreed with that in Fig. (2.3.4) for the rectangular x-ray exposure.

This above model has been extended to consider the three-dimensional configuration in Fig. (2.3.8) of mask, proximity gap, resist, and substrate (Dai and Nassar, 1998a). The heat equations in expressions (2.3.4) and (2.3.5) are extended, at steady state, to include mask and gap which gives for mask

$$-k_1 \left(\frac{\partial^2 T_1}{\partial x^2} + \frac{\partial^2 T_1}{\partial y^2} + \frac{\partial^2 T_1}{\partial z^2} \right) = g_1(x, y, z), \qquad (2.3.25a)$$

gap

$$-k_2 \left(\frac{\partial^2 T_2}{\partial x^2} + \frac{\partial^2 T_2}{\partial y^2} + \frac{\partial^2 T_2}{\partial z^2} \right) = g_2(x, y, z), \qquad (2.3.25b)$$

resist

$$-k_3 \left(\frac{\partial^2 T_3}{\partial x^2} + \frac{\partial^2 T_3}{\partial y^2} + \frac{\partial^2 T_3}{\partial z^2} \right) = g_3(x, y, z), \qquad (2.3.25c)$$

and substrate

$$-k_4 \left(\frac{\partial^2 T_4}{\partial x^2} + \frac{\partial^2 T_4}{\partial y^2} + \frac{\partial^2 T_4}{\partial z^2} \right) = g_4(x, y, z). \qquad (2.3.25d)$$

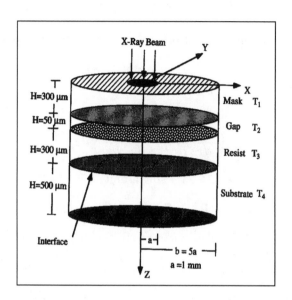

Fig. 2.3.8. Three-dimensional circular configuration of mask, gap, resist, and substrate (Reproduced from Dai and Nassar 1998a by permission of Taylor & Francis Inc., http://www.routledge-ny.com)

The boundary conditions are now as follows. On the top surface of the mask $(z = 0)$,

$$k_1 \frac{\partial T_1}{\partial z} = h_c(T_1 - T_\infty).$$

On the bottom surface of the mask $(z = H_1)$,

$$-k_1 \frac{\partial T_1}{\partial z} = -k_2 \frac{\partial T_2}{\partial z} \quad T_1 = T_2.$$

On the top surface of the resist $(z = H_1 + H_2)$,

$$-k_2 \frac{\partial T_2}{\partial z} = -k_3 \frac{\partial T_3}{\partial z} \quad T_3 = T_2,$$

and on the bottom surface of the resist $(z = H_1 + H_2 + H_3)$,

$$-k_3 \frac{\partial T_3}{\partial z} = -k_4 \frac{\partial T_4}{\partial z} \qquad T_3 = T_4.$$

On the side walls and bottom surface of the substrate, it is assumed that the temperature T_i $(i = 1, 2, 3, 4)$ for each of mask, gap, resist and substrate is equal to the ambient temperature T_∞.

Consider the steady state heat equation for layer i

$$-k \left(\frac{\partial^2 T_i}{\partial x^2} + \frac{\partial^2 T_i}{\partial y^2} + \frac{\partial^2 T_i}{\partial z^2} \right) = g_i. \tag{2.3.26}$$

Following the same finite element procedure, one obtains the steady state form of Eq. (2.3.19)

$$k K \mathbf{T} - k D \frac{\partial^2 \mathbf{T}}{\partial z^2} = D \mathbf{f} \tag{2.3.27a}$$

where the element of K is now

$$k_{qp} = \int \int \left(\frac{\partial \varphi_p}{\partial x} \frac{\partial \varphi_q}{\partial x} + \frac{\partial \varphi_p}{\partial y} \frac{\partial \varphi_q}{\partial y} \right) dx dy. \tag{2.3.27b}$$

A discretized form of Eq. (2.3.26) using a finite difference scheme gives

$$k K \mathbf{T}_m - \lambda D (\mathbf{T}_{m+1} - 2\mathbf{T}_m + \mathbf{T}_{m-1}) = D \mathbf{f}_m \qquad m = 1, ..., N_z, \tag{2.3.28}$$

where $\mathbf{T}_m = \mathbf{T}_m(m\Delta z)$ and $\lambda = k/\Delta z^2$. To simplify the system, a preconditioned Richardson iteration is applied (Canuto et al. 1988; Deville et al. 1994) with regard to Eq. (2.3.27) to obtain for each layer i :

$$k D_1 (\mathbf{T}_i)_m^{(n+1)} - \lambda D \left((\mathbf{T}_i)_{m+1}^{(n+1)} - 2(\mathbf{T}_i)_m^{(n+1)} + (\mathbf{T}_i)_{m-1}^{(n+1)} \right)$$
$$= k D_1 (\mathbf{T}_i)_m^{(n)} - \lambda D \left((\mathbf{T}_i)_{m+1}^{(n)} - 2(\mathbf{T}_i)_m^{(n)} + (\mathbf{T}_i)_{m-1}^{(n)} \right)$$
$$- \beta \left[k K (\mathbf{T}_i)_m^{(n)} - \lambda D \left((\mathbf{T}_i)_{m+1}^{(n)} - 2(\mathbf{T}_i)_m^{(n)} + (\mathbf{T}_i)_{m-1}^{(n)} \right) - D \mathbf{f}_m \right]$$

$$m = 1, ..., N_z \qquad n = 1, 2, 3, ... \tag{2.3.29}$$

where D_1 is a diagonal matrix with a diagonal element, $d_i = 2k_{ii}$, k_{ii} is the entry on the main diagonal line of the matrix K, and $i = 1, 2, ..., N$. Here, β is a relaxation parameter. The preconditioned Richardson iteration has the advantage of fast convergence in the microscale case.

Equation (2.3.29) can now be applied to solving the heat equations in Eq. (2.3.25). At each iteration step, $(\mathbf{T}_i)_m^{(n+1)}$ is assumed to satisfy the discretized boundary conditions for Eq. (2.3.25). On the walls and bottom surface of the substrate, $(\mathbf{T}_i)_m^{(n+1)} = T_\infty$ $(i = 1, 2, 3, 4)$. For the computation, the Gaussian elimination procedure is used to solve the four block tridiagonal linear systems resulting from the preconditioned Richardson iteration in Eq. (2.3.29). Parameters used for conductivity $(k, W/cm/K)$ were 2.0, .00152, .00198, and

1.5 for Berylium, He gas (100 mbar), PMMA, and silicon, respectively. Parameters used for I_0 (W/cm^2) were 2.042, 0.0, 1.823, and 1.424, respectively. For μ (cm^{-1}), the values used were 40.81, 0.0, 50.26, and 99.42, respectively. (Weast 1985; Barron 1985; Brandup and Immergut 1989).

The radius of the exposed circular area in the mask was taken as $r = 1mm$, and the convection coefficient was, as before, $h_c = 60\ W/m^2/K$. Figure (2.3.9) presents the temperature rise along the z-axis at the center of the exposed circular pattern. It is seen that the maximum temperature rise was 5.72 K and that it occurred in the resist which is in agreement with previous results from two layers (resist and substrate) for the same parameter values and for circular or rectangular geometry.

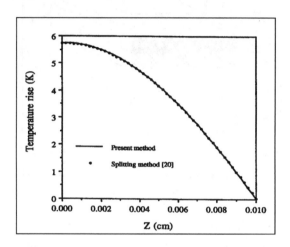

Fig. 2.3.9. Temperature profiles along the z-axis. Maximum temperature rise is in the resist (Reproduced from Dai and Nassar 1998a by permission of Taylor & Francis Inc., http://www.routledge-ny.com)

A treatment of the three-dimensional case for four layers (mask, proximity gap, resist, and substrate) where the exposure area is rectangular in shape as seen in Fig. (2.3.10) was considered by Dai and Nassar (1998b) for the steady state case. The governing heat conduction equation is expressed as

$$-k_i \left(\frac{\partial^2 T_i}{\partial x^2} + \frac{\partial^2 T_i}{\partial y^2} + \frac{\partial^2 T_i}{\partial z^2} \right) + g_i(x, y, z, t) \quad i = 1, 2, 3, 4 \quad (2.3.30)$$

where the index i refers to the layers (mask, gap, resist, and substrate). The boundary conditions at the interface between layers are

$$-k_i \frac{\partial T_i}{z} = -k_{i+1} \frac{\partial T_i}{\partial z} \quad T_i = T_{i+1} \quad i = 1, 2, 3. \quad (2.3.31)$$

These boundary conditions imply continuity of heat flux at the interface as well as continuity of temperature caused by a perfect thermal contact. At the upper surface of the mask, the convective boundary condition is

$$-k_1 \frac{\partial T_1}{\partial z} = h_c(T_1 - T_\infty) \tag{2.3.32}$$

and $T_4 = T_\infty = 300K$ at the bottom surface of the substrate, and $T_i = T_\infty = 300K$ at the side walls.

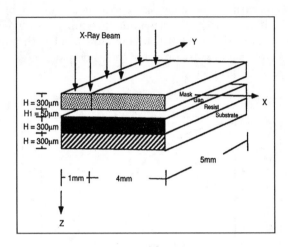

Fig. 2.3.10. Schematic diagram of an x-ray irradiated process (Reproduced from Dai and Nassar 1998b by permission of MCB University Press)

A solution to Eq. (2.3.30) was obtained using finite difference and applying the preconditioned Richardson method and the parallel Gaussian elimination technique. Details of the methods are in Dai and Nassar (1998b). Numerical results were obtained where the mask, gap, resist, and substrate were beryllium, low pressure helium gas, PMMA, and silicon, respectively. The values for the parameter k $(W/cm/K)$ for the four layers were 1.98×10^{-3}, 1.5, 2.01, and 0.152 for PMMA, silicon, mask, and gap, respectively. The source term is given by

$$g = I_0\mu \exp(-\mu z); \quad 0 \le x \le 1\,mm \quad 0 \le y \le 5\,mm.$$

The intensity of the beam was $I_0 = 3.4\ W/cm^2$ and μ was $\frac{10^4}{106cm}$. The convective heat transfer coefficient was $h_c = 60\ W/m^2/K$. The temperature rise profile at $x = 0.7\ mm$ (where the maximum temperature occurred) and $y = 2.5mm$ as seen in Fig. (2.3.11) was in agreement with the two layer case (resist and substrate) in Dai and Nassar (1997b). The temperature profile over mask, gap, resist, and substrate at $x = 0.7\ mm$ and $y = 2.5\ mm$ is presented in Fig. (2.3.11). It is seen for this example that the rise is mostly in the resist.

The same four-layer model for the configuration in Fig.(2.3.10) and for the unsteady state situation was considered by Hegab et al. (1998). In this case, the heat equation is expressed as

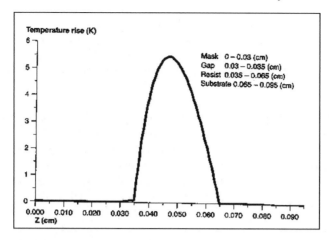

Fig. 2.3.11. Temperature profile along the vertical line at x = 0.7mm and y = 2.5mm in the mask, gap, resist, and substrate (Reproduced from Dai and Nassar 1998b by permission of MCB University Press)

$$\rho_i c_i \frac{\partial T_i}{\partial t} = k_i \left(\frac{\partial^2 T_i}{\partial x^2} + \frac{\partial^2 T_i}{\partial y^2} + \frac{\partial^2 T_i}{\partial z^2} \right) + g_i(x, y, z, t) \qquad i = 1, 2, 3, 4. \quad (2.3.33)$$

The heat source is expressed as

$$g_i(x, y, z, t) = I_{0i} \mu_i \exp \left[-\mu_i \left(d - z - \sum_{j=1}^{i-1} z_j \right) \right] \qquad 0 \le x \le a, 0 \le y \le b$$

$$(2.3.34)$$

$$= 0 \qquad \text{otherwise,}$$

where I_{oi} is the irradiation intensity and μ_i is the absorption coefficient in the *ith* layer. Here, z is in the opposite direction from that in Figure (2.3.10) where $z = 0$ is the bottom of the substrate. The boundary conditions at the interfaces is given by Eq. (2.3.31). At the upper surface of the mask the convective boundary condition is the same as that in Eq. (2.3.32). At the bottom surface of the substrate and at the side walls, the temperature is set at $300K$. A solution to Eq. (2.3.33) was obtained using finite difference and applying the three-dimensional Douglas alternating direction implicit (ADI) scheme (Douglas 1962) and the parallel Gaussian elimination technique. Numerical results were obtained where the mask, gap, resist, and substrate were berylium, low pressure helium gas, PMMA, and silicon, respectively. The values for the parameters k, μ, and I_0 were the same as those in Dai and Nassar (1998a). In addition, the values for the parameter $\rho \, (g/cm^3)$ were $1.845, 1.603 \times 10^{-5}, 1.18$, and

2.33 for berylium, helium gas, PMMA, and silicon, respectively. Also, the corresponding values for c_p (kJ/kgK) were 1.825, 5.2, 1.42, and 0.715. The convective heat transfer coefficient was $h_c = 60\ W/m^2K$, and the layer thicknesses were 300 μm, 50 μm, 300 μm, and 500 μm for mask, gap, resist, and substrate, respectively. In Fig. (2.3.10), the width of the exposed area is $a = 4\ mm$ with the layers taken as squares of dimension $b = 2\ cm$. One beam scan was considered with a velocity of 5 cm/s. The temperature rise profile at $x = 0.4\ cm$ and $y = 1.0\ cm$ for various grid sizes is presented in Fig. (2.3.12). The profile is similar to that in Fig. (2.3.11) in that the highest temperature rise is in the resist. However, the maximum temperature rise is less than that in Fig. (2.3.11). Because the parameters and layer thicknesses are the same in both models, this difference in temperature rise may be attributed to scanning.

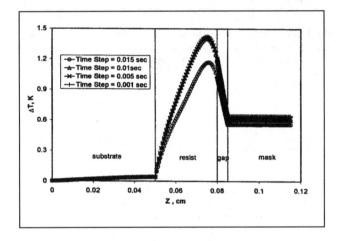

Fig. 2.3.12. Temperature rise profile at the middle of beam (x = 0.4cm and y = 1.0 cm for various time steps (Reproduced from Hegab et al. 1998 by permission of Taylor & Francis Inc., http://www.routledge-ny.com)

The four-layer model in the unsteady state condition has been considered when the geometry of the exposed area was circular (Dai and Nassar, 1999). The hybrid finite element finite difference scheme was used to obtain a solution for the temperature gradient. The finite element method is that presented in Dai and Nassar (1997b), extended to include the gap and mask. A finite difference Crank-Nicolson type scheme was used for discretization. The Gaussian elimination scheme was used to solve the resulting block tridiagonal linear system of equations. Computational details are presented in Dai and Nassar (1997b). A numerical example based on the configuration in Fig. (2.3.8) and assuming the same parameter values for k, W, μ, ρ, and c_p as in Hegab et al. (1998) was considered. Results of transient temperature profiles along the z-axis where the maximum rise occurred are presented in Figure (2.3.13). It is seen, as expected, that the temperature rise increases with time until steady state is reached. Steady state is reached rapidly and the temperature rise at steady state is in agreement with that obtained in Dai and Nassar (1998a) using the preconditioned Richardson method.

These above four layer models, with circular and rectangular exposure patterns, are useful in predicting the effect of x-ray lithography operating parameters such as beam scan speed, material properties, layer thickness, gap distance, and gas type and pressure on the temperature profile, and ultimately on displacements or distortions. As such, they can be used for optimizing an x-ray lithography process. The hybrid finite element finite difference method is general and can be applied to analyzing configurations where the exposed area is of arbitrary geometry.

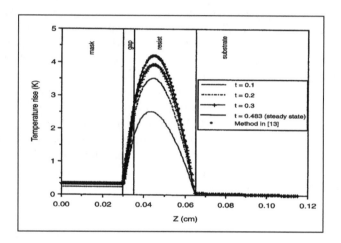

Fig. 2.3.13. Transient temperature profiles along the z-axis. Method in [13] is from Dai and Nassar 1998a (Reproduced from Dai and Nassar 1999a by permission of MCB University Press)

2.4 Microstructure Transfer Accuracy

In deep x-ray lithography, accuracy of the microstructure can be limited by secondary radiative processes, namely diffraction of radiation, radiation transmission through an insufficiently thick absorber, divergence of the synchrotron radiation, energy deposition through emission of photoelectrons and Auger electrons, fluorescence, and photon scattering. All of these factors can cause exposure of the resist below the absorber region of the mask which can lead to inaccuracy in the sidewall profile. A simulation study by Feiertag et al. (1997) investigated the effects of the above factors on the accuracy of the sidewall profile in the resist. Fresnel diffraction theory with superimposed complex radiation amplitudes (Heinrich et al. 1981) were used to study the effects of diffraction (due to the irradiation of the absorber edge) and partial transmission of radiation through the absorber on the irradiation intensity distribution in the resist near the absorber edge. From the irradiation intensity distribution in the x (lateral) and y (depth) directions and the 1.8 kJ cm^{-3} dissolution limit for the PMMA resist, it was possible to predict the resist sidewall profile after development. Deviations of the wall profile from an ideal vertical line profile was less than 0.1 μm.

Divergence of irradiation below the absorber edge is given by the angular deviation of the light emitting electrons of the storage ring and the emission angle distribution of the synchrotron radiation. Electron and synchrotron radiation angular deviations can be approximated by a Gaussian distribution with variance equal to the sum of the variances of the electron divergence and the irradiation divergence. The effect of divergence is expected to be small since synchrotron radiation is parallel. Calculation of wall profile based on the Gaussian distribution indicated small deviations from an ideal vertical line. The intensity distribution of diffraction was convoluted with the energy distribution caused by divergence to give the combined intensity distribution.

When an incident x-ray photon is absorbed by an atom in the resist, the energy transfer from the photon to a core electron of the atom can cause the emission of a photoelectron. The kinetic energy of the photoelectron is the photon energy minus the binding energy of the core electron. The excited atom relaxes either by emitting a secondary electron (Auger electron) or a photon (Fluorescence). Emitted electrons undergo elastic and inelastic scattering in the vicinity of the point of absorption which can lead to energy dissipation. A Monte Carlo procedure was used in order to determine the impact of these electrons on energy deposition below the absorber edge. The steps involved photoabsorption, electron scattering, and energy dissipation and were modeled in the same manner as described in Murata (1985).

The probability that an atom interacts with a photon is given by

$$p_k = \frac{C_k \sigma_k}{\sum_i C_i \sigma_i}, \tag{2.4.1}$$

where C_k is the weight fraction of atoms k in the resist and σ_k is the corresponding total absorption cross section. For a photon energy greater than the binding energy of the K-shell, the total absorption cross section can be expressed as

$$\sigma_{tot} = \sigma_K + \sigma_L, \tag{2.4.2}$$

where σ_K and σ_L are the K-shell and L-shell cross sections. Hence, the probability of a photon being absorbed by the K-shell is given by

$$p_K = \frac{\sigma_K}{\sigma_{tot}}. \tag{2.4.3}$$

For oxygen and carbon atoms, it was assumed that when a K-shell photoelectron is generated, the excited atom generates an Auger electron. The angular probability distribution function (pdf) of the emitted photoelectron is given by

$$f(\theta) = 0.75 \sin^3(\theta), \tag{2.4.4}$$

where θ is the angle between the incident photon and the ejected electron. The pdf for the Auger electron is given by

$$f(\theta) = .5 \sin \theta. \tag{2.4.5}$$

Elastic scattering of the electrons in the resist was modeled using the screened Rutherford equation

$$\frac{d\sigma}{d\Omega} = \frac{Z(Z+1)e^4}{4E^2(1 - \cos\theta + 2\beta)^2}, \tag{2.4.6}$$

where β is the effective screening parameter, Z is the atomic number, E is the electron energy in keV, and e is the electric charge. The energy loss by collisions along an electron trajectory is modeled based on the Bethe equation (Murata, 1985) which is given by

$$\frac{dE}{ds} = -\frac{2\pi e^4 nZ}{E} \ln \frac{1.166E}{J}; \quad E \geq 6.338J, \tag{2.4.7}$$

and

$$\frac{dE}{ds} = -\frac{2\pi e^4 nZ}{1.26\sqrt{JE}}; \quad E < 6.338J, \tag{2.4.8}$$

Where J is the mean ionization potential in keV, and n is the number of atoms in a unit volume. The length of the path is determined from the energy loss equation. From the above equations one can determine by simulation the paths of electrons in the resist and at the same time the energy distribution along these paths. The predicted wall profile (for resists more than $10\mu m$ in thickness) based on a dissolution limit of $1.8 \ kJcm^{-3}$ indicated a larger deviation from the ideal vertical line than was the case for diffraction and divergence.

Energy deposition caused by fluorescence photons was simulated by first generating a primary photon and determining its path length (l) in the resist from the expression

$$l = \frac{-\ln R}{\mu(E)}, \tag{2.4.9}$$

where R is a uniform random number between 0 and 1 and $\mu(E)$ is the absorption coefficient of the resist. If the path length is less than the thickness of the resist, then the photon is absorbed in the resist and its energy recorded. When the photon path length is larger than the thickness of the resist, the photon enters the substrate. The path length of the photon in then determined to find the absorption point in the substrate. If the primary photon energy

at the point of absorption in the substrate is larger than the K-shell binding
energy, a fluorescence photon is generated and its angle and path length determined. For a fluorescence photon that reaches the resist, a path length is
calculated in order to determine the point of absorption. The energy deposited
at that point is recorded and the process is repeated with the generation of
another primary photon. For fluorescence to be of significance the material
must be made of high atomic number elements. In addition, since the energy
of photons in deep x-ray lithography is between 1-10 keV, only materials with
atoms having K-shell binding energies in this range are likely to be a source
for fluorescence. As such, materials like silicon, titanium, nickel or copper are
likely candidates for fluorescence emission. Figure (2.4.1) gives a distribution
of the dose in $kJ\ cm^{-3}$ deposited by fluorescence radiation from a titanium
substrate. It is seen that significant damage to the resist in the vicinity of the
substrate can occur as a result of fluorescence.

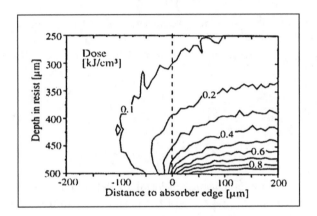

Fig. 2.4.1. Distribution of dose deposited in the resist (500 μm thick) by
fluorescence radiation from a titanium substrate (reprinted from Feiertag et al
1997 by permission of IOP Publishing Limited)

A Monte Carlo approach similar to the one applied in the case of fluorescence was used to determine the effect of scattering in the mask membrane,
resist and substrate. Calculations showed that less than 120 $J\ cm^{-3}$ was deposited in the resist below the absorber as a result of scattered radiation from
a beryllium membrane (500 μm). Scattering in a 500 μm thick PMMA resist
resulted in a dose deposition of 100 $J\ cm^{-3}$ or less. The effect of scattering in
the titanium substrate resulted in a deposition of less than 10 $J\ cm^{-3}$ below
the absorber. Hence, scattering has negligible effect since the sum of all three
contributions was less than the dissolution limit.

In conclusion the study showed that photoelectron emission and fluorescence are more important than diffraction, beam divergence, or scattering
and can contribute to inaccuracies in wall profiles. Comparisons between cal-

culated (based on contributions from diffraction, absorber transmission, beam divergence, photo- and Auger electrons, and fluorescence) and experimental results of wall profile were performed for titanium, copper, and carbon substrates with a 500 μm PMMA resist. The worst case scenario was that for titanium which is presented in Fig. (2.4.2). The curvature in the resist edge near the substrate is attributed to fluorescence radiation emitted from the substrate as shown in Fig.(2.4.1). The curvature near the surface of the resist was attributed to the developer and/or thermal deformation of the mask. In general there was fair agreement between experimental and calculated results.

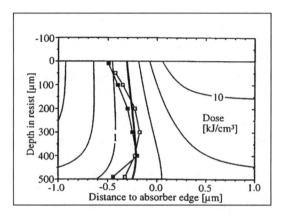

Fig. 2.4.2. Comparison of calculated dose contour (bold line, 1.8kJcm^{-3}) and experimental measurements (squares) for a titanium substrate. Contour lines represent calculated dose distribution due to diffraction, absorber transmission, beam divergence, fluorescence, photo- and Auger-electrons (reprinted from Feiertag et al 1997 by permission of IOP Publishing Limited)

Griffiths and Ting (2001, 2002) studied the effect of x-ray fluorescence emission in the substrate of the PMMA resist on resist sidewall profile. Their model is based on the one-dimensional multiwave model discussed below (Griffiths et al. 1999), which describes primary exposure resulting from x-ray transmission (through beam filters, mask, and resist), coupled with development. A portion of the primary dose absorbed in the substrate gives rise through fluorescence to a secondary dose in the resist. Effect of the secondary dose on the sidewall profile was examined for the situation where the mask consisted of a 100 μm silicon membrane with a 15 μm gold absorber and where the PMMA resist was 1 mm in thickness. The primary top- and bottom-surface doses in the open region of the PMMA were 8.7 and 5 kJ/cm^3, respectively. Also, the same top and bottom doses under the absorber (or masked region) were 0.11 and 0.088 kJ/cm^3, respectively. Results were in general agreement with those in Feiertag et al. (1997) in that fluorescence caused the wall profile in the resist near the substrate surface to encroach a certain distance under the absorber edge. This distance, however, varied with the type of substrate. For silicon and nickel substrates, the distance or deviation after 5 hours of development time at 25^0C was 20 nm and 2 μm, respectively. The deviation for nickel increased to 10 μm at 35^0C. While deviations in the wall profile for silicon were limited to the region near the substrate surface, that for nickel extended to a height of about 100 μm. These deviations in the sidewall for nickel are substantial and larger in magnitude than was reported by Feiertag et al. (1997) or by Zumaque et al. (1997) which was in the submicron range. Excessive sidewall dissolutions were found also for copper and titanium. On the other hand, sidewall dissolutions for aluminum, carbon, and berylium (materials low in atomic number) were low yielding sidewall tolerances less than 0.01 μm. Also, deviations in the sidewall profile were found to depend on the feature geometry, the magnitude of the primary dose, and the development temperature. It was concluded that a 20% increase in the primary dose increased the sidewall dissolution by about a factor of two. Also, a similar increase in sidewall disolution was caused by an increase of 5^0C in the development temperature.

A simulation study to investigate the effects of Fluorescence photon and photo-and Auger-electron emission in the mask membrane, absorber, and resist on the accuracy of the developed microstructure has been reported by Zumaque et al. (1997). The simulation techniques were similar to those in Murata (1985) and in Feiertag et al. (1997). The x-ray lithography experimental set up consisted of synchrotron radiation passing through the collimator, and the vacuum window before reaching the mask membrane. For the purpose of simulation, the number of photons, N, generated per unit wavelength λ and solid angle Ω per second was expressed as

$$N(\lambda, \Omega) = \frac{3e^2}{8\pi^2 h r^2} \gamma^2 \left(\frac{\lambda_c}{2\lambda}\right)^3 (1+\gamma^2\theta^2)^2 \left[K_{2/3}^2(\varepsilon) + \frac{\gamma^2\theta^2}{1+\gamma^2\theta^2}K_{1/3}^2(\varepsilon)\right],$$

(2.4.10)

where e is the electron charge, h is the Planck's constant, r is the radius of the storage ring, θ is the emission angle of the photon relative to the plane of the beam, $\gamma = E/m_e c^2$, $\lambda_c = \frac{4\pi r}{3\gamma^3}$, $\varepsilon = (1+\gamma^2\theta^2)^{3/2}\frac{\lambda_c}{2\lambda}$, and K_v are the modified Bessel functions. In order to determine the number of photons that pass through the collimator, Eq. (2.4.10) is integrated over the solid angle ψ (centered in the beam plane) defined by the collimator to give

$$N(\lambda, \psi) = \int_{\Omega_\psi} N(\lambda, \Omega) d\Omega.$$

(2.4.11)

The probability density of a photon passing through a vacuum window of thickness W is given by

$$p(\lambda, \psi, W) = \frac{\mu(\lambda)e^{-\mu(\lambda)W}N(\lambda, \psi)}{\int_0^\infty N(x, \psi)dx},$$

(2.4.12)

where $\mu(\lambda)$ is the absorption coefficient in the window. From the above considerations, the probability that a primary photon of wavelength λ reaches the mask membrane is given by

$$P(\lambda, \psi, W) = \int_0^\lambda p(x, \psi, W)dx. \tag{2.4.13}$$

Once the photon is generated based on the above probability its absorption position, z, is determined at random based on the exponential distribution

$$p(z) = \mu_m(\lambda)e^{-\mu_m(\lambda)z}, \tag{2.4.14}$$

where μ_m is the absorption coefficient in the mask membrane. If z is larger than the thickness of the membrane, then the photon is positioned at the membrane absorber interface and a new z value is generated from the exponential distribution using the absorption coefficient for the absorber. If z happens to exceed the thickness of the absorber, the process is repeated until the photon is absorbed in one of the layers, resist or substrate, or it exits through the substrate. Once the primary photon is absorbed at a given z location, a photoelectron and an Auger electron or a fluorescence photon are generated using similar techniques as that described in Murata(1985) and Feiertag et al. (1997). Results of the energy density contours with depth, which relate to wall profiles in the resist, show that the contours slope gradually inward (towards the center of the absorber hole) from top to bottom. However, near the surface of the substrate the contours bend outward away from the center of the absorber hole. This result is in agreement with results in Feiertag et al. (1997). Lateral degradation (assuming a dissolution limit of $1.5kJ\ cm^{-3}$) was most noticeable in the resist near the surface of the substrate, but was in the submicrometer range and did not explain the degradation observed in experiments which is of the order of micrometers. This pattern degradation above the substrate surface can be eliminated by using a substrate with low atomic number. It was concluded that in order to obtain a more accurate prediction of the wall profile, one must simulate the resist development process.

Resist development plays an important role in sidewall taper. As the sidewalls begin to form during development, they become exposed to the developer. Since small doses are encountered in the masked region as a result of some x-ray transmission through the absorber, the walls in contact with the developer begin to recede laterally into the masked area of the resist. The sidewall inward taper is caused by the fact that the sidewall contact time with the developer decreases from top to bottom of the resist. Increasing the absorber thickness reduces x-ray transmission to the masked region, which in turn reduces sidewall taper. A numerical model was developed by Griffith et al. (1999) to simulate the exposure and development process. The model describes the local absorbed dose in the exposed and masked regions of the resist, the kinetics of the development rate, and the transport of resist fragments from the dissolution surface to the top of the resist. Using this model, one may optimize exposure (by optimizing the absorber thickness, beam filter thickness, and exposure time) subject to a constraint with regard to sidewall taper. Dose rate in the resist (PMMA) is modeled by considering wavelength-dependent transmission and absorption cross-section. Secondary effects, such as photo- and Auger-electrons and fluorescence, were not considered in the model. With these considerations, the transmitted beam power, $p_{0,k}$, is given by

$$p_{0,k} = p_{i,k}e^{-\rho\sigma_{t,k}l}, \tag{2.4.15}$$

where $p_{i,k}$ is the incident beam power at photon energy E_k, ρ is material density, l is material thickness, and $\sigma_{t,k}$ is the transmission cross-section of the material at photon energy E_k. The beam passes through one or more filters (layers) before reaching the absorber layer. Hence, power transmission can be computed sequentially (power transmission from one layer is used as incident power for the next layer) using Eq. (2.4.15). The local dose rate, q_k, in the PMMA resist depends on the transmitted power and the absorption cross-section for some photon energy E_k and is given by

$$q_k = \rho\sigma_{t,k}p_{0,k}. \tag{2.4.16}$$

The total dose rate, $\frac{dQ}{dt}$, (energy per unit time per unit volume) is then obtained as

$$\frac{dQ}{dt} = \sum_k q_k\delta E_k, \tag{2.4.17}$$

where δE_k represents one-half of the band of energy between E_{k-1} and E_{k+1}. For a constant incident power, the total dose Q (kJ/cm^3) is then the total dose rate multiplied by the exposure time t. It is clear that as the filter thickness increases, only higher energy photons are transmitted. High energy photons give a more uniform transmitted power through the thickness of the resist. Hence, by adjusting the thickness of the filters in the model one can obtain a desired top to bottom dose ratio in the resist. Also, by adjusting the absorber thickness one can obtain the desired top or bottom dose in the masked region of the resist.

For modeling the development process, the dissolution rate was assumed to depend on the reaction kinetics and on fragment transport away from the dissolution surface. As such, the linear development rate in the y-direction (resist depth) can be expressed as

$$\frac{dy}{dt} = cU_0, \tag{2.4.18}$$

where U_0 is the development rate for a given temperature and dose and c is the solvent volume fraction at the dissolution surface. Here,

$$c = \frac{ShD}{ShD + U_0y}, \tag{2.4.19}$$

where Sh is the Sherwood number (or ratio of the convective to diffusive transport rates) and D is the diffusivity coefficient of the PMMA fragment. This coefficient was approximated as

$$D = D_0 \left(\frac{w_m}{w_e}\right)^2 \sqrt{\frac{T}{T_0}}, \qquad (2.4.20)$$

where D_0 is the monomer diffusivity ($D_0 \approx 0.4 \times 10^{-9} m^2 s^{-1}$) at $T_0 = 293K$, w_m is the monomer molecular weight of PMMA ($w_m = 100 \ g/mol$), and w_e is the fragment molecular weight.

The lateral development rate at the surface of the resist in the masked region is not diffusion limited. Hence, it can be assumed to depend only on the reaction rate. As such, the lateral rate of wall dissolution can be expressed as

$$\frac{dx}{dt} = U_0. \qquad (2.4.21)$$

The kinetic-limited development rate is related to the Arrhenius equation and is given by

$$U_0 = a\frac{(Q/b)^c}{1 + (Q/b)^c} \exp\left[-\frac{E_a}{R}\left(\frac{1}{T} - \frac{1}{T_r}\right)\right], \qquad (2.4.22)$$

where c, b , and a are development rate constants (For the ALS-1.9 Gev synchrotron source, $a = 12.15 \ \mu m/\min$, $b = 4.47 \ kJ/cm^3$, and $c = 3.8$), $T_r = 308K$, and the activation energy is given by

$$E_a = \frac{\alpha}{1 + (Q/\beta)^\kappa}. \qquad (2.4.23)$$

Here, $\alpha = 139 \ kJ/mol$, $\beta = 8.32 \ kJ/cm^3$, and $\kappa = 2.38$. Model results were presented for prescribed sidewall tolerance, structure size, resist thickness, mask thickness, and development temperature. Figure (2.4.3) presents the development time, the minimum thickness of the gold absorber, the top dose of the resist, and the masked dose as a function of the bottom dose for different values of the sidewall dissolution distance, ε ($\varepsilon/2 =$ the lateral distance from the edge of the absorber to the sidewall in the masked region at the top of the resist). Here, the filter is beryllium ($127 \ \mu m$), the mask substrate is silicon ($100 \ \mu m$), the resist height is 1 mm, the absorber hole width is d=10 mm and the development temperature is $35^0 C$. It is seen, as expected, that the minimum absorber thickness increases with a decrease in the sidewall dissolution distance. For instance, at a bottom dose of 4 kJ/cm^3 the minimum absorber thickness increases from 5.5 μm at $\varepsilon = 10$ um to 9.7 μm at $\varepsilon = 0.1$ μm. Also, for the same sidewall dissolution distance, increasing the bottom dose necessitates increasing the minimum absorber thickness. Development time and top surface dose do not change with absorber thickness or with sidewall distance since the height and width of the structure as well as the transmitted power source are fixed. As expected, development time increases as the bottom dose decreases. Other results showed that for fixed dissolution distance ($\varepsilon = 1$) the development time increased significantly with a decrease in the feature width at the bottom surface of the resist (substrate surface). The increase in development time is due to the diffusion-limited transport in high aspect ratio features. In order to hold the dissolution distance constant, it was necessary to increase the minimum thickness of the absorber so as to reduce the dose in the masked region of the resist. In order not to increase the minimum absorber thickness, the development time must be reduced. This result can be accomplished by enhancing the diffusion-limited transport rate through acoustic agitation.

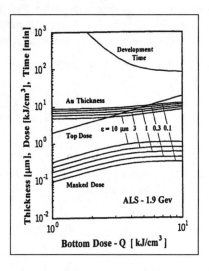

Fig. 2.4.3. Development time, minimum gold absorber thickness, resist top dose, and masked dose as a function of resist bottom dose for different values of the sidewall dissolution distance, ε. The development temperature is T=35^{0}C, resist height is 1mm, width of absorber hole is 10mm, beam beryllium filter is 127 μm thick , and the silicon mask substrate is100 μm thick (reprinted from Griffiths et al. 1999 by permission of IOP Publishing Limited)

Transport limitations between the top surface of the resist mold and the dissolution surface during development can cause, in addition to sidewall development into lateral areas in the masked region, incomplete dissolution of the exposed resist, undercutting of the resist-substrate interface, and lengthy development time. For electroplating, transport limitations can cause nonuniform deposition rates and poor metal morphology . A promising means of enhancing the transport rate in an attempt to remedy these problems is acoustic agitation which has been studied by Nilson and Griffiths (2000, 2002) and Nilson et al. (2001). Figure (2.4.4) is a schematic diagram of the geometry of the mold that has been considered. The bath fluid above the mold is assumed to be well mixed with a uniform flow in the direction of the arrows. Acoustic field of a given frequency and intensity causes high frequency pressure variations at the mouth of the mold which drive the acoustic motion within the mold.

To determine the governing equations based on the Navier Stokes equations, the pressure, p, density, ρ, and velocity $\mathbf{u} = ui + vj$ were approximated by the series expansions

$$\mathbf{u} \approx \mathbf{u}_a + \mathbf{u}_s \qquad p \approx p_0 + p_a + p_s \qquad \rho \approx \rho_0 + \rho_a + \rho_s, \qquad (2.4.24)$$

where the acoustic terms with the subscript a are harmonic functions of time and those with the subscript s are steady acoustic streaming terms independent of time. The terms p_0 and ρ_0 are constant. Replacing the velocity, density, and pressure in the Navier Stokes equations by their linear forms from Eq. (2.4.24) and equating terms of like order, one obtains the linearized harmonic acoustic equations given by

$$\frac{\partial \rho_a}{\partial t} + \rho_0 \nabla \cdot \mathbf{u}_a = 0 \qquad \rho_0 \frac{\partial \mathbf{u}_a}{\partial t} = -\nabla p_a + \mu \nabla^2 \mathbf{u}_a + (\mu_b + \frac{\mu}{3}) \nabla (\nabla \cdot \mathbf{u}_a), \quad (2.4.25)$$

where μ and μ_b are the shear and bulk velocities. The acoustic pressure can be expressed as $p_a = c_0^2 \rho_a$, where c_0 is the sound velocity. Equations governing the steady flow are obtained by averaging the series expansion of the Navier Stokes equations over time (Nyborg 1998) to give

$$\nabla \cdot \mathbf{u}_s = 0 \qquad \nabla p_s - \mu \nabla^2 \mathbf{u}_s = F \qquad F = -\rho_0 \langle (\mathbf{u}_a \cdot \nabla) \mathbf{u}_a + \mathbf{u}_a (\nabla \cdot u_a) \rangle, \qquad (2.4.26)$$

where the force F arises from the Reynolds stresses as a consequence of the harmonic motion. One may solve Eq.(2.4.25) for $\mathbf{u}_a$ and use the solution in Eq. (2.4.26) to obtain $\mathbf{u}_s$. The velocity of the steady flow is then used to solve for the concentration C, of the PMMA fragment in resist development or of the metal ion concentration in electroplating, from the steady state diffusion-convective equation

$$\nabla \cdot (\mathbf{u}_s C) = \nabla \cdot (D \nabla C), \qquad (2.4.27)$$

where D is the diffusivity coefficient of the fragment or metal ion.

Equation (2.4.26) was rewritten in terms of the stream function and vorticity and solved numerically along with Eq. (2.4.27) using the second-order finite difference scheme on a rectangular mesh over the whole region in Fig. (2.4.4). The resulting equations contained three dimensionless parameters, the aspect ratio (h/w), the ratio of the acoustic boundary layer to the mold width $(\frac{\delta}{w})$, and the Peclet number, calculated as $\frac{u_{s0} w}{D}$, where u_{s0} is the streaming speed adjacent to a solid surface calculated as $u_{s0} = \frac{I}{\rho c^2}$, where I is the acoustic intensity, ρ is the liquid density, and c is the speed of sound.

The Sherwood number, Sh, which represents the ratio of the convective to diffusive transport rates in the vertical direction, was computed from the expression

$$Sh = \left(\frac{h}{w}\right) \frac{1}{D\Delta C} \int_0^w D\frac{\partial C}{\partial y} dx. \qquad (2.4.28)$$

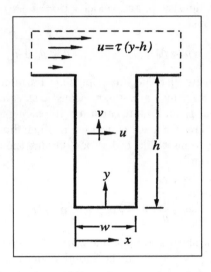

Fig. 2.4.4. Schematic of a LIGA Mold used in the model (reprinted from Nilson and Griffiths 2000 by permission of SPIE)

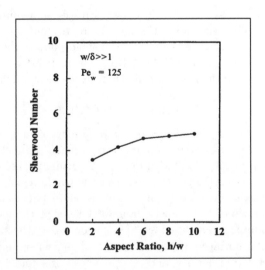

Fig. 2.4.5. Sherwood number as a function of the aspect ratio (reprinted from Nilson and Griffiths 2000 by permission of SPIE)

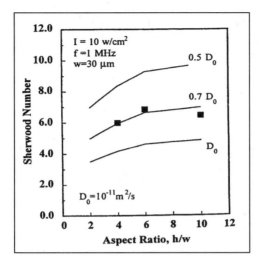

Fig. 2.4.6. Comparison of the calculated Sherwood number (solid line) with experimental data (from Zanghellini et al. 1998 by permission of Springer Verlag) for different values of the aspect ratio (reprinted from Nilson and Griffiths 2000 by permission of SPIE)

Results from the model revealed that the steady flow through the mold as a result of acoustic agitation was toroidal with downflow along the walls and upflow in the center of the cavity. The maximum flow velocity near the wall was about one-fourth of u_{s0}. For $\rho = 1000 \ kg/m^3$, $c = 1500 \ n/s$, and $I = 10 \ W/cm^2$, the downflow speed near the wall is about $11 \ \mu m/s$. Numerical results in Fig. (2.4.5) show that the Sherwood number (at a Peclet number of 125) increased from about 3.5 to 4.5 with an increase in the aspect ratio, h/w, of the mold from 2 to about 10. These Sh values indicate that acoustic agitation is beneficial in enhancing transport and that its value does not diminish with an increase in the aspect ratio. Further results showed that the Sherwood number (a measure of development rate) increased with the Peclet number. Hence, to increase the development rate one may increase the Peclet number by increasing the streaming speed u_{s0} (which is proportional to the bath intensity). For a fixed bath intensity or streaming speed, one can increase the Peclet number and development rate by increasing the width of the mold (and height also if the aspect ratio is held constant). Comparison of calculated Sh (solid lines) to experimental observations (measured as the ratio of the development rates with and without agitation) is presented in Fig. (2.4.6) for a bath intensity, $I = 10W/cm^2$, at a frequency of 1 MHz. The width of the mold was 30 μm. The three calculated lines were for three different fragment diffusivities, $0.5D_0$, $0.7D_0$, and D_0, where D_0 is the monomer diffusivity and equals $10^{-11}m^2/s$. Best agreement with observed enhancement of development rate occurs for a fragment diffusivity of $0.7 \times 10^{-11} \ m^2/s$. This value of

$0.7\ D_0$ seems reasonable since fragment diffusivity is expected to be less than D_0. Further work along these lines, in conjunction with experimentation, is needed for process optimization.

3

Laser Chemical Vapor Deposition

3.1 Introduction

Laser chemical vapor deposition (LCVD) is a process by which an organometallic compound in the form of a precursor gas is induced, by light or heat from a laser source, to deposit in the form of a solid on a substrate. Deposition can occur through a photolytic or pyrolytic mechanism, or a combination of both mechanisms. In photolytic LCVD, deposition occurs when the precursor gas is dissociated directly by light. The dissociation can occur over the surface (when the laser beam is parallel to the surface) or at the surface (when the beam is incident to the surface as in direct write photolysis). The deposition rate in photolytic LCVD is very low (usually in Angstroms/unit time) and is feasible for depositing thin films without heating the substrate. It has been employed for the manufacture and repair of integrated circuit interconnects (Osgood 1983; Baum 1992).

In pyrolytic LCVD, the precursor gas is induced to deposit on the surface through laser-induced surface heating of the substrate. The deposition rate is much higher than that of photolytic LCVD, which makes it better suited as an important technique in freeform fabrication of high aspect ratio microstructures. While pyrolytic LCVD has been widely explored for depositing thin metallic films, the deposition of three-dimensional microstructures has only recently become of widespread interest. The LCVD process is computer controlled and is capable of producing microstructures of arbitrary shapes. Furthermore, the process is rapid, flexible, and relatively inexpensive to operate. The pyrolytic LCVD set-up consists of a laser, a vacuum chamber, and a moveable target as shown schematically in Fig. (3.1.1). The laser beam is focused through a chamber window onto a substrate target. Scanning of the substrate surface by the laser beam is accomplished through a computer-controlled movement of the stage in the $x - y$ plane. A precursor gas is introduced into the chamber where it reacts at or near the focal spot of the laser on the substrate, leaving behind a solid deposit. Precursors are chosen so that the by-products of the reaction are volatile and return to the surrounding gas mixture (Maxwell 1996). Using this technique, a solid can be deposited on a spot in the form of a fiber or rod. By scanning the beam in one or in two dimensions in the $x-y$ plane, solids can be deposited to form two-dimensional or three-dimensional microstructures of various shapes.

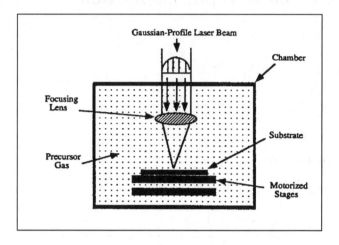

Fig. 3.1.1. LCVD schematic diagram (reprinted from Dai et al 1999b by permission of Taylor & Francis, Inc., http://www.routledge-ny.com)

Determining the rate of deposition on the substrate surface is crucial for modeling the process and for assessing its feasibility for the fabrication of microstructures. The deposition rate in pyrolytic LCVD is influenced by the surface temperature, by the reaction activation energy, and by the rate of transport of the precursor gas and by-product molecules to and from the surface. The deposition rate is kinetically limited when it depends only on the activation energy and surface temperature. On the other hand, the rate is mass transport limited when the reaction is controlled by the diffusion of the precursor and by-products to and from the surface.

3.2 Kinetically Limited Deposition

For a kinetically limited LCVD process, the local deposition rate is determined by the Arrhenius equation

$$R_n = k_0 e^{-E_a/R_g T} \tag{3.2.1}$$

where R_n is the local rate of deposition in the normal direction to the surface, k_0 is a rate constant, E_a is the activation energy in kcal/mol, R_g is the universal gas constant, and T is the local surface temperature (Baeuerle 1990).

It is seen from this equation that the rate of deposit is determined by the activation energy of the precursor and by the surface temperature. The rate increases exponentially at first with an increase in temperature and then levels off. This behavior is explained by the fact that as the temperature rises the

rate of conversion of the precursor gas increases until a point in temperature is reached where the ability of the precursor gas to arrive at the surface is not sufficient to maintain the ongoing conversion rate and the deposition rate levels.

There are several models developed for kinetically limited pyrolytic LCVD to simulate growth or deposition on a spot, on a strip through laser scanning of a line, and on a surface through scanning in the $x - y$ direction. A one-dimensional model to simulate the growth of a stripe on a semi-infinite substrate using laser direct writing in pyrolytic LCVD was developed by Arnold et al. (1993a). For the steady state case, the temperature distribution induced by the laser can be obtained from the energy balance equation expressed as

$$F\kappa_D \frac{\partial^2 T_D}{\partial x^2} - \kappa_s \int_{-r_D}^{r_D} \frac{\partial T_s}{\partial z} \Big|_{z=0} dy + P_a \delta x = 0, \qquad (3.2.2)$$

where F is the cross-section of the stripe; κ_D and κ_s are the thermal conductivities of the deposit and substrate, respectively; T_D and T_s are the temperature of the deposit and substrate, respectively; $P_a \ \delta x$ is the absorbed laser power per length δx; and r_D is equal to half the width of the stripe. Here, the beam is scanned along a line in the x-direction where the center of the beam of width ω is at the origin ($x = 0$). The case is considered where the ratio of the thermal conductivity of the deposit to that of the substrate is much larger than one $\left(\kappa^* = \frac{\kappa_D}{\kappa_s} \gg 1 \right)$. Since $\kappa^* \gg 1$, the temperature at the surface of the substrate $T_s(z = 0)$ can be approximated by T_D. Approximating $\frac{\partial T_s}{\partial z}$ by $\frac{\Delta T_D}{r_D}$, where $\Delta T_D = (T_D - T(\infty))$, the integral in Eq. (3.2.2) becomes $\eta \kappa_s \Delta T_D$, where η is approximately equal to 2. From these approximations, Eq. (3.2.2) may be expressed as

$$l^2 \frac{\partial^2 T_D}{\partial x^2} - (T_D - T(\infty)) + \frac{P_a \delta x}{\eta \kappa_s} = 0, \qquad (3.2.3)$$

where $l^2 = F\kappa^*/\eta$ and $T(\infty) = T(x \to -\infty)$. The boundary condition at $x = a$ (the forward edge of the stripe) is $\frac{\partial T_D}{\partial x} \big|_{x=a} = 0$ and that at x infinity is $T(\infty)$. The solution to Eq. (3.2.3) for $x < 0$ is

$$T_D(x) = T(\infty) + \Delta T_c \exp(x/l), \qquad (3.2.4)$$

where

$$\Delta T_c \equiv \Delta T(x = 0) = \frac{P_a}{2\eta l \kappa_s} \left(1 + \exp(-\frac{2a}{l}) \right), \qquad (3.2.5)$$

and for $0 < x < a$

$$T_D(x) = T(\infty) + \Delta T_e \cosh(\frac{x-a}{l}), \tag{3.2.6a}$$

where

$$\Delta T_e = \Delta T(x=a) = \frac{P_a}{\eta l \kappa_s} \exp(-\frac{a}{l}). \tag{3.2.6b}$$

For estimation purposes, the cross-section of the stripe is approximated by

$$F \approx \zeta h r_D, \tag{3.2.7}$$

where h is height of the stripe behind the laser beam and ζ is a coefficient depending on the geometry of the cross-section of the stripe. Likewise, r_D is approximated by ξa where ξ is of order one.

For the simulation of growth through deposition at $x = 0$, the height of the stripe h behind the laser beam is given by the expression

$$W(T_c) - v_s \frac{h}{\gamma a} = 0, \tag{3.2.8}$$

where $W(T_c)$ is the growth rate from the Arrhenius equation for T_c, the temperature at the center of the laser beam ($x = 0$), v_s is the scan velocity of the beam in the x-direction, and γ is a coefficient of order one. From Eqs. (3.2.4) and (3.2.6a), one obtains

$$\mu = \frac{a}{l} = \mathrm{arccos}\, h\left(\frac{\Delta T_c}{\Delta T_{th}}\right), \tag{3.2.9}$$

where $\Delta T_{th} = T_{th} - T(\infty)$ and $T_{th} = \Delta T_e + T(\infty) =$ threshold temperature below which no deposition takes place. From Eqs. (3.2.6a), one obtains

$$l = \frac{P_a}{\eta \kappa_s \Delta T_{th}} \exp(-\mu), \tag{3.2.10}$$

which when combined with Eq. (3.2.9) gives

$$r_D = \xi a = \xi l \mu = \frac{\xi P_a}{\eta \kappa_s \Delta T_{th}} \mu \exp(-\mu). \tag{3.2.11}$$

From Eq. (3.2.7) and the relations $r_D = \xi a$ and $l^2 = \frac{F \kappa^*}{\eta}$, one has the relation

$$h = \frac{\eta}{\zeta \xi \kappa^*} \frac{a}{\mu^2} = \frac{P_a}{\zeta \xi \kappa_D \Delta T_{th}} \mu^{-1} \exp(-\mu). \tag{3.2.12}$$

From Eqs. (3.2.8) and (3.2.12), one obtains

$$v_s = \frac{\gamma \zeta \xi}{\eta} \kappa^* \mu^2 W(T_c). \tag{3.2.13}$$

From Eqs. (3.2.9) and (3.2.11)-(3.2.13), one may calculate the height (h) and width ($2r_D$), as a function of v_s, the scanning velocity.

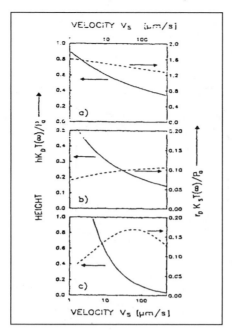

Fig. 3.2.1. Normalized width and height of a stripe as related to the scan velocity, v_S (reprinted from Arnold et al. 1993a with permission from Elsevier Science)

It is seen from Eqs. (3.2.11) and (3.2.12) that the width and height of a stripe increases linearly with the absorbed laser power, P_a. This is in agreement with experimental observations from different systems (Baeuerle 1986). Figure (3.2.1) presents the relationship, as calculated from Eqs. (3.2.11)-(3.2.13), between the laser scan velocity, v_s and the normalized height and width, $\frac{h\kappa_D T(\infty)}{P_a}$ and $\frac{r_D \kappa_s T(\infty)}{P_a}$, respectively. Parameters employed were (a) $k_0 = 6.6 \times 10^{12} \mu m/s$, $E_a/k_B T(\infty) = 45$, $T_{th}/T(\infty) = 1.17$, $\kappa^* = 30$; (b) $k_0 = 6.6 \times 10^9 \mu m/s$, $E_a/k_B T(\infty) = 90$, $T_{th}/T(\infty) = 3.7$, $\kappa^* = 15$; and (c) $k_0 = 16.05 \ \mu m/s$, $E_a/k_B T(\infty) = 5.7$, $T_{th}/T(\infty) = 2.7$, $\kappa^* = 17$. For all cases $\eta = 1.6$, $\zeta = 4/3$, $\xi = 1.25$, and $\gamma = 1.3$. It is seen from the figure that the height decreases monotonically with an increase in scan velocity. However, the width can increase or decrease monotonically, or it can increase to a maximum and then decrease with an increase in the scan velocity depending on the parameters employed. Figure (3.2.2) presents experimental results on width and height of W stripes as a function of scan velocity for two mixtures of $WCl_6 + H_2$. The pressure of H_2 was 50 $mbar$. The parameters were $P(514.5 \ nm \ Ar^+) = 645 \ mW$, $\omega_0 = 7.5 \ \mu m$, absorptivity = 0.55, $T(\infty) = 443K$, $E_a/k_B T(\infty) = 5.7$, $\kappa^* = 17$, $\kappa_s = 0.032 \ W \ cm^{-1} K^{-1}$; $\eta = 1.6$, $\zeta = 4/3$, $\xi = 1.25$, $\gamma = 1.3$, (a) $p(WCl_6) = 0.49 \ mbar$, $k_0 = 7.15 \mu m/s$, $T_{th}/T(\infty) = 2.4$; (b) $p(WCl_6) = 1.1 \ mbar$, $k_0 = 16.05 \mu m/s$, $T_{th}/T(\infty) = 2.7$.

The figure shows good agreement between experimental results on height and width of W stripes as a function of scan velocity and calculated results (solid and dashed curves) from Eqs. (3.2.11)-(3.2.13).

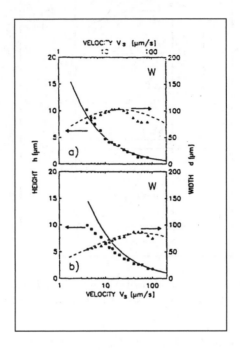

Fig. 3.2.2. Normalized width and height as related to the scan velocity, v_S, for two mixtures of WCl_6+H_2 (reprinted from Arnold et al. 1993a with permission from Elsevier Science)

A two-dimensional model to simulate the deposition of spots and direct writing of lines in pyrolytic LCVD has been presented by Arnold and Baeuerle (1993b). As in Arnold et al. (1993a), the substrate is assumed to be semi-infinite, and the origin of the coordinate system on the surface coincides with the center of the beam. The heat equation can be used for calculating the temperature distribution on the surface of deposit induced by the laser beam. This may be expressed in three dimensions as

$$c_D \rho_D \frac{\partial T_D}{\partial t} = \nabla_3 \left(\kappa_D(T_D) \nabla_3 T_D \right) + Q\left(x, y, z, t \right), \qquad (3.2.14)$$

with boundary conditions

$$\kappa_D(T_D) \frac{\partial T_D}{\partial z} \mid_{z=0} = J_{loss}(z=0),$$

and

$$-\kappa_D(T_D)\frac{\partial T_D}{\partial \hat{\eta}}\mid_{z=h}= J_{loss}(z = h),$$

at the substrate surface and the deposit surface, respectively. Here, c_D is the specific heat, ρ_D is the mass density of the deposit, κ_D is the thermal conductivity in the deposit, Q is the heat source term, J_{loss} is the heat loss at the substrate and deposit surfaces, T_D is the surface temperature of the deposit, and $\hat{\eta}$ is the unit normal to the surface $z = h(x,y)$. For convenience, Eq.(3.2.14) was transformed by integrating it over the region of the deposit $0 \leq z \leq h$. This gives

$$\int_0^h c_D\rho_D\frac{\partial T_D}{\partial t}dz = \int_0^h \frac{\partial}{\partial z}\left(\kappa_D(T_D)\frac{\partial T_D}{\partial z}\right)dz +$$

$$\int_0^h \nabla_2\left(\kappa_D(T_D)\nabla_2 T_D\right)dz$$

$$+ \int_0^h Q\left(x,y,z,t\right)dz. \qquad (3.2.15)$$

After some approximations, namely replacing $\frac{\partial T_D}{\partial \hat{\eta}}$ by $\frac{\partial T_D}{\partial z} - \Delta_2 h.\Delta_2 T_D$ in the case of a flat deposit and equating $T_D(x,y,z,t)$ to $T_D(x,y,0,t)$ which holds when the thermal conductivity of the deposit is substantially larger than that of the substrate, or strictly speaking, if $(h/r_D\kappa^* \ll 1)$, it is seen that the integral in Eq. (3.2.15) becomes

$$c_D\rho_D\frac{\partial T_D}{\partial t} = -J_{loss}(z = h) - J_{loss}(z = 0) + \nabla_2\left(h\kappa_D(T_D)\nabla_2 T_D\right)$$

$$+ \int_0^h Q\left(x,y,z,t\right)dz. \qquad (3.2.16)$$

To obtain a solution for Eq. (3.2.16) the following assumptions were made: (1) $J_{loss}(z = h) = 0$. Also, since $T_s = T_D\mid_{z=0}$, $J_{loss}(z = 0) = \kappa_s(T_s)\frac{\partial T_s}{\partial z}\mid_{z=0}$. (2) the temperature is at steady state implying that $\frac{\partial T_D}{\partial t} = 0$. (3) the laser light is totally absorbed in which case $\int_0^h Q\left(x,y,z,t\right)dz = I_a(x,y) = AI(x,y)$, where A is the absorptivity. Based on these assumptions, Eq. (3.2.16) becomes

$$\kappa_s(T(\infty))\frac{\partial \theta_s}{\partial z}\mid_{z=0}= I_a(x,y) + \kappa_D(T(\infty))\Delta_2\left(h(x,y)\Delta_2\theta_D\right), \qquad (3.2.17)$$

where θ_D and θ_s are the linearized temperatures for deposit and substrate using the Kirchoff transform (Baeuerle 2000). Equation (3.2.17) may be considered as a boundary condition at $z = 0$ for the three-dimensional heat equation in the substrate,

$$\nabla_3^2\theta_s = 0. \qquad (3.2.18)$$

Using the Green's function technique, Eq. (3.2.18) was solved to give

$$\theta_s = \theta_s^0 + \frac{\kappa^*(T(\infty))}{2\pi}\left\{\nabla_2(h\nabla_2\theta_D) * \frac{1}{\mid r \mid}\right\}, \qquad (3.2.19a)$$

where

$$\theta_s^0 = \frac{1}{2\pi\kappa_s(T(\infty))}\left\{I_a * \frac{1}{|r|}\right\}, \qquad (3.2.19b)$$

and r is the radial distance from the beam center. The equation describing the shape of a deposit in one dimension is expressed as

$$\frac{\partial h}{\partial t} = W(T_D) + v_s\frac{\partial h}{\partial x}, \qquad (3.2.20a)$$

where $W(T_D)$ is the growth rate, expressed by an Arrhenius type equation

$$W(T_D) = k_0\exp(-\frac{E_a}{R_g T_D})\left[1 + \exp\left(\frac{T_{th} - T_D}{\delta T_{th}}\right)\right]^{-1}, \qquad (3.2.20b)$$

where T_{th} is the threshold temperature and δT_{th} its width. Here, $h = 0$ is the boundary condition as well as the initial condition for Eq.(3.2.20a).

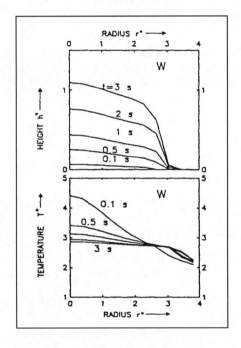

Fig. 3.2.3. Normalized surface temperature and height of Tungsten spot deposits at various times of growth as a function of the normalized radial distance from the center of the laser beam (reprinted from Arnold et al. 1993b with permission of Elsevier Science)

From numerical solutions to Eqs. (3.2.19) and (3.2.20), the model was applied to simulate the deposition of spots and direct writing of lines. Figure (3.2.3) shows simulation results for the growth of Tungsten (W) deposited on a quartz (SiO_2) substrate. The kinetic parameters employed were $E_a/R_gT(\infty) = 5.68$, $K_0/\omega_0 = 2.14$, $T_{th}/T(\infty) = 2.71$, $\frac{\delta T_{th}}{T(\infty)} = 0.01$, $\kappa^*(T(\infty)) = 50.56$, $AI_0\omega_0/[T(\infty)\kappa_s(T(\infty))] = 10.3$. The thermal conductivity of W was expressed as $\kappa_D = c_1 + c_2/T - c_3/T^2$, where $c_1 = 42.65 Wm^{-1}K^{-1}$, $c_2 = 1.898 \times 10^4 Wm^{-1}$, and $c_3 = 1.498 \times 10^6 Wm^{-1}K$. Conductivity for the silicone substrate was approximated by $\kappa_s = a_1 + a_2T$, where $a_1 = 0.9094 Wm^{-1}K^{-1}$ and $a_2 = 1.422 \times 10^{-3} Wm^{-1}K^{-2}$. It is seen that, for different stages of growth, the height and surface temperature of the spot decrease with an increase in the radial distance from the center of the beam. The simulation of the deposition of nickel (Ni) is shown in Fig. (3.2.4). Here, $\kappa^*(T(\infty) = 300K) = 30$, $k_0/\omega_0 = 1.1x10^{13}$, $E_a/R_gT(\infty) = 45$, $AI_0\omega_0/[T(\infty)\kappa_s(T(\infty))] = 1.33$, $T_{th}/T(\infty) = 1$ (no threshold). Because no temperature threshold was assumed, it is seen that there was no abrupt termination in lateral growth as was the case with tungsten. As for W, the height and surface temperature of the deposit decreased with an increase in the lateral distance from the beam center. With regard to direct writing of lines, it was observed that the simulated shape of stripes deposited from $WCl_6 + H_2$ fluctuated or remained uniform depending on the parameters employed in the model. The above results are in qualitative agreement with experimental observations (Baeuerle 1986).

A one-dimensional model describing the growth of a fiber or rod in pyrolytic LCVD has been developed by Arnold et al. (1996). A schematic diagram of the deposition process is shown in Fig. (3.2.5). It is assumed that deposition takes place primarily in the tip region where the beam is focused and where the temperature is high. Changes in the radius of the rod occur mainly in this region. In the region below the deposition region (below the dashed line in the diagram), growth does not occur, and the radius is constant. Heat is conducted through the rod into this lower region where it is gradually lost to the surrounding gas. In this model, the temperature gradient in the radial direction is assumed to be negligible. Hence, temperature change occurs only in the z or axial direction of the rod. The stationary heat equation for the rod, taking into account the change in the radius as a function of z and heat loss to the gas phase, may be expressed as

$$\frac{\partial}{\partial z}(\pi R^2 \kappa_D \frac{\partial T_D}{\partial z}) + 2\pi R \kappa_g \frac{\partial T_D}{\partial z}\ |_{r=R} = 0, \qquad (3.2.21)$$

with boundary conditions

$$-(\pi R^2 \kappa_D \frac{\partial T_D}{\partial z})\ |_{z=0} = AP$$

and

$$T_d\ |_{z=\infty} = T\infty.$$

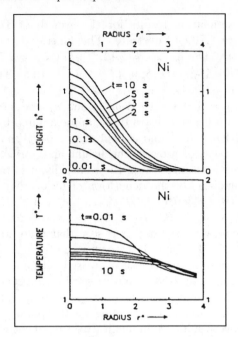

Fig. 3.2.4. Normalized surface temperature and height of Nickel spot deposits at various times of growth as a function of the normalized radial distance from the center of the laser beam (reprinted from Arnold et al. 1993b with permission of Elsevier Science)

Here, $R = R(z)$, κ_D and κ_g are the thermal conductivities for deposit and gas, which depend in general on the temperature of the deposit, T_D, and the temperature of the gas, T_g, respectively. The ambient temperature is denoted by $T(\infty)$, and A is the absorptivity coefficient.

Growth, caused by the deposition of the precursor gas, is in the direction normal to the surface at each point $(z, R(z))$ (Fig. 3.2.5). Under equilibrium conditions, the change in the radius with z can be written as

$$\frac{dR}{dz} = \frac{W(z)}{W(0)} \left(1 + (\frac{dR}{dz})^2 \right)^{1/2}, \qquad (3.2.22)$$

with initial condition $R\mid_{z=0}= c w_0$, c being a constant. Here, $W(z)$ is the growth rate at z and is expressed as

$$W(z) = k_0 \exp(-E_a/R_g T(z)). \qquad (3.2.23)$$

As the surface temperature becomes high, the reaction rate proceeds quickly and may become influenced by the diffusion of the precursor gas to the reaction site on the surface. With high reaction rates, the growth rate is approximated by the equation

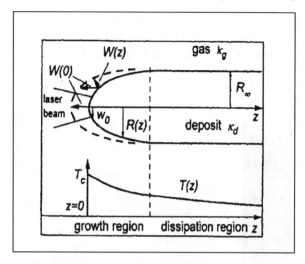

Fig. 3.2.5. Schematic diagram of a rod deposition process showing the shape of the rod R(z) and the temperature distribution T(z) as a function of z. In the dissipation region, R(z) is a constant and T(z) approaches T_0 (reprinted from Arnold et al. 1996 Fig. 1 with permission of Springer-Verlag)

$$W = \frac{W(z)}{1 + \frac{\rho_d W(z) R}{\rho_g D}}, \qquad (3.2.24)$$

where D is the gas diffusion coefficient and ρ_D and ρ_g are the densities of the solid and precursor gas, respectively.

Equations (3.2.21) and (3.2.22) can be solved numerically in order to determine the shape or radius of the growing rod. Instead, the Kirckhoff transform was applied to Eq. (3.2.21) in order to exclude the dependencies of thermal conductivities on temperature in the equation, and a numerical solution was obtained on the transformed equation. An approximate analytical solution was obtained when gas diffusion was ignored. From model calculations, ignoring any gas diffusion limitation, it was determined, as expected, that the radius of the rod as well as the temperature increased with an increase in laser power. An increase in laser spot size (or radius of the laser beam, ω_0) caused the temperature at the tip of the rod to decrease and the radius to increase. An increase in the activation temperature, T_a, causes an increase in the temperature at the tip of the rod, $T(z = 0)$, and a decrease in its radius. As expected, the temperature at the tip increases with a decrease in the thermal conductivity. For low laser power, as the thermal conductivity of the solid increases relative to that of the gas, the radius of the rod increases. When growth is affected by gas diffusion, the effect leads to a decrease in the temperature at the tip and to an increase in the radius. The effect of gas diffusion becomes more pronounced with temperature-dependent thermal conductivities. A fit of the model to experimental data on radius size concerning the growth of Si rods from SiH_4 was satisfactory in the range of medium laser power. The model was not satisfactory for low or high laser powers. Nevertheless, this one-dimensional simple approach is useful in understanding observed experimental results.

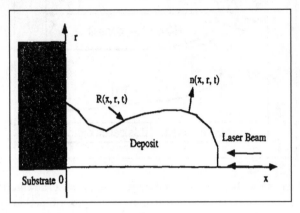

Fig. 3.2.6. Schematic diagram of an axisymmetric rod (reprinted from Dai et al 1999b by permission of Taylor & Francis, Inc., http://www.routledge-ny.com)

A one-dimensional axisymmetric model for simulating rod growth in pyrolytic LCVD has been discussed by Maxwell (1996) and by Dai et al. (1999b). Consider an axisymmetric rod growth as shown schematically in Fig. (3.2.6). The surface of the deposit is defined by the vector function $R(x, r, t)$. Growth through deposition is in the normal direction to the surface, and its magnitude for kinetically limited reaction rates may be expressed by the Arrhenius equation

$$\mathbf{R}(x, r, t) = k_0 \exp(-E_a/R_g T_d(x, r, t))\hat{\eta}, \qquad (3.2.25)$$

where $\hat{\eta} = \hat{\eta}(x, r, t)$ is the unit normal to the deposit surface at point (x, r) at time t. If the radius of the deposit can be expressed as a function of x, then $R(x, r, t) = R(x, t)$, and the deposit surface and normal vector can be expressed as

$$R(x, t) = x(t) \cdot \hat{x} + r(x, t) \cdot \hat{r}, \qquad (3.2.26)$$

and

$$\hat{\eta}(x, t) = -\frac{1}{\psi} \left(\frac{dr(x, t)}{dx} \right) \cdot \hat{x} + \frac{1}{\psi} \cdot \hat{r}, \qquad (3.2.27)$$

where

$$\psi = \sqrt{1 + \left(\frac{dr(x, t)}{dx} \right)^2}.$$

From the above considerations, Eq. (3.2.25) becomes

$$\mathbf{R}(x,t) = k_0 \exp(-E_a/R_g T_d(x,t)) \left[-\frac{1}{\psi} \left(\frac{dr(x,t)}{dx} \right) \cdot \hat{x} + \frac{1}{\psi} \cdot \hat{r} \right]. \quad (3.2.28)$$

To solve for the shape of the rod, the surface temperature $T_d(x,t)$ in Eq. (3.2.28) must be determined. In principle, the surface temperature may be obtained from the heat equations for deposit and substrate, namely

$$c_d \rho_d \frac{\partial T_d}{\partial t} = \nabla(\kappa_d \nabla T_d) + Q_{in} - Q_{loss} \quad (3.2.29)$$

and

$$c_s \rho_s \frac{\partial T_s}{\partial t} = \nabla(\kappa_s \nabla T_s). \quad (3.2.30)$$

The interfacial equations between deposit and substrate are

$$T_d = T_s \qquad \text{and} \qquad \kappa_d \frac{\partial T_d}{\partial \hat{\eta}} = \kappa_s \frac{\partial T_s}{\partial \hat{\eta}}.$$

However, one may simplify these equations (Maxwell 1996) by assuming that (1) the heat flux in and out of the deposit is in equilibrium and the temperature distribution is at steady state; (2) material properties are constants and the influence of the substrate is not important, especially for long rods; (3) the temperature gradient is largely along the x-direction. Let γ be the deposit volume per unit surface area, then Q_{in} and Q_{loss} are related to the heat fluxes, $Q_{in}^{''}$ and $Q^{''}$, by the relation $\gamma Q = Q^{''}$. With these assumptions, Eq. (3.2.28) simplifies to

$$\kappa_d \frac{d}{dx} \left(\pi r^2(x) \frac{d\theta}{dx} \right) + 2\pi r(x) Q_{in}^{''} - 2\pi r(x) Q_{loss}^{''} = 0, \quad (3.2.31)$$

where $\theta = T_d - T_\infty$ (ambient temperature). Heat loss to convection and radiation from the surface may be expressed as

$$Q_{loss}^{''}(x,r) = h_{conv}\theta + \varepsilon_s \sigma \theta^4, \quad (3.2.32)$$

where h_{conv} is the heat transfer coefficient and the term $\varepsilon_s \sigma \theta^4$ represents heat loss through radiation. The heat source from a Gaussian beam may be expressed as

$$Q_{in}^{''}(x,r) = 2P \frac{\Lambda}{\pi \omega^2 (x-z)} e^{-2r^2/\omega^2(x-z)} \left[\hat{x} \cdot \hat{\eta}(x) \right], \quad (3.2.33)$$

where P is the laser power, Λ is absorptivity, and $\omega(x-z)$ is the laser spot radius at any axial position $(x-z)$, where z is the focal position relative to the rod tip. The laser input flux may be considered as boundary condition at the tip of the rod. This boundary condition may be expressed as

$$\frac{d\theta}{dx} = \frac{1}{\pi \kappa_d r^2} (\lambda P - h_{conv}\theta), \quad (3.2.34)$$

$$\frac{d\theta}{dx} = \frac{1}{\pi \kappa_d r^2} Q_{cond},$$

(3.2.35)

Where x is at the tip and $\lambda = \frac{2\Lambda}{\pi \omega^2}$. The boundary condition at the base of the rod may be represented by where x is at the base of the rod. Heat passing into the substrate generates an average temperature at the interface between the rod and the substrate. This temperature may be expressed for a disc on a semi-infinite substrate as (Carslaw and Jaeger, 1959)

$$\theta = \frac{8r}{3\pi \kappa_s} Q_{cond}.$$

(3.2.36)

As such, the boundary condition at the base becomes

$$\frac{d\theta}{dx} = \frac{3\kappa_s \theta}{8\kappa_d r^3}.$$

(3.2.37)

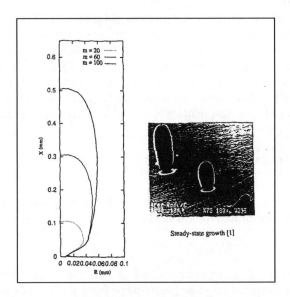

Fig. 3.2.7. Simulated and grown rod deposited from nickel on a graphite substrate (reprinted from Dai et al. 1999b by permission of Taylor & Francis, Inc., http://www.routledge-ny.com)

Equations (3.2.31)-(3.2.34) and (3.2.37) coupled with Eqs. (3.2.26)-(3.2.28) may be solved to determine the evolution of the shape of the deposit in the axial and radial directions. A finite difference procedure is described in Dai et al. (1999b) to solve the equations. A numerical example is provided simulating the deposition of Nickel on a graphite substrate. Parameters employed are $E_a = 9.4 \times 10^4 J/mol$, $R_g = 8.314 J/(mol\ K)$, $k_0 = 1.37 \times 10^4 mm/s$, $\kappa_d =$

$6.55 \times 10^{-2} W/(mm\ K)$, $\kappa_s = 1.7 \times 10^{-3} W/(mm\ K)$, $T_\infty = 300K$, $P = 0.1W$, $\omega = 0.01\ mm$, $A = 1$, and

$$h_{conv}(x) = \frac{Nu\kappa_{gas}}{2r(x)},$$

where Nu (Nusselt number) $= 0.36$ and $\kappa_{gas} = 0.001$.

Figure (3.2.7) shows good qualitative agreement between a simulated rod growth deposited from nickel on a graphite substrate and an experimentally grown rod (Maxwell 1996) employing the parameters listed above. In this simulation, thermal conductivities were assumed to be constants. When the thermal conductivity is a function of temperature, Eqs. (3.3.35)-(3.3.37) become (McMordie 1962)

$$\frac{d\theta_d}{dx} = \frac{1}{\pi r^2} Q_{cond},$$

$$\theta_s = \frac{8r}{3\pi} Q_{cond},$$

and

$$\frac{d\theta_d}{dx} = \frac{3\theta s}{8r^3},$$

where $\theta_d = \kappa_d - \kappa_d(\infty)(T_\infty - T_0)$ and $\theta_s = \kappa_s - \kappa_s(\infty)(T_\infty - T_0)$.

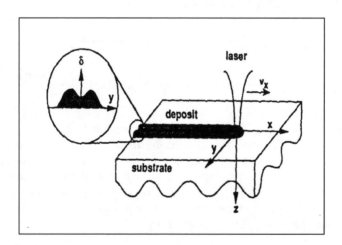

Fig. 3.2.8. Schematic description of the modeled LCVD process. The origin is located at the laser-beam center and moves with the laser at a scan velocity v_x (reprinted from Han and Jensen 1994 with permission from the American Institute of Physics)

A one-dimensional model to simulate deposition of copper on silicon and metal substrates through line scanning was developed by Han and Jensen

(1994). The model was used to explain the volcano-like cross-sections often observed in the deposition of metal stripes. Like other models discussed so far, this model considers heat transfer and chemical reaction and takes into account the effect of the growing deposit on the surface temperature distribution. The model is formulated to specifically address the deposition of copper from the organometallic source, copper(I)-hexafluoroacetylacetonate-trimethylvinylsilane (Cu(hfac)tmvs). The model is based on the schematic diagram in Fig. (3.2.8). The origin of the coordinate system is located at the center of the laser beam and moves with the beam at a scan velocity v_x in the x direction. The temperature distribution on the surface may be obtained from the heat equations for deposit and substrate. For unsteady state conditions, these equations are given by

$$\rho_d c_d \frac{\partial T_d}{\partial t} = -\rho_d c_d v_x \frac{\partial T_d}{\partial x} + \nabla \left(\kappa_d \nabla T_d \right) + Q_d, \tag{3.2.38}$$

and

$$\rho_s c_s \frac{\partial T_s}{\partial t} = -\rho_s c_s v_x \frac{\partial T_s}{\partial x} + \nabla \left(\kappa_s \nabla T_s \right) + Q_s, \tag{3.2.39}$$

where s and d refer to substrate and deposit, respectively, T_s and T_d are the temperatures, κ_s, κ_d are temperature-dependent conductivities, c is specific heat, and ρ is the density. For a Gaussian beam energy distribution, the heat sources, Q_d and Q_s may be expressed as

$$Q_d = \frac{\alpha_d P(1 - R)}{\pi \omega^2} \exp(-\alpha_d z) \exp\left(-\frac{2(x^2 + y^2)}{\omega^2}\right),$$

$$Q_s = \frac{\alpha_s P(1 - R)}{\pi \omega^2} \exp(-\alpha_d \delta) \exp(-\alpha_s z) \exp\left(-\frac{2(x^2 + y^2)}{\omega^2}\right),$$

where P is the laser power, $\alpha's$ are the absorptivity coefficients by each layer, ω is the laser spot radius, R is the reflectivity coefficient, and $\delta = \delta(x, y)$ is the deposit thickness. The boundary conditions are $T_d = T_s =$ room temperature when far away from the laser beam (i.e, $x \to \pm\infty$, $y \to \infty$, and $z \to \infty$); $\hat{\eta} \nabla T_d = 0$ or $\hat{\eta} \nabla T_s = 0$ on the surface ($z = -\delta$); and, at the interface ($z = 0$), $T_d = T_s$, and $\kappa_d \nabla T_d = \kappa_s \nabla T_s$. The initial condition on the substrate surface at $t = 0$ is $T_d = T_s =$ room temperature. The equation describing the evolution of the deposit thickness, $\delta(x, y)$, is given by

$$\frac{\partial \delta}{\partial t} + v_x \frac{\partial \delta}{\partial x} = r_z, \tag{3.2.40}$$

where r_z is the surface reaction or growth rate. The initial condition is $\delta(x, y, z) = 0$ at $t = 0$, and the boundary condition is $\delta \to 0$ as $x \to \infty$.

The reaction mechanism for copper deposition involves three steps. In the first step, $Cu(hfac)tmvs$ in the gas phase is absorbed (with rate constant k_1) on the surface and dissociates to give $Cu(hfac)$ and $tmvs$, which is desorbed into the gas phase. In the second step, $Cu(hfac)$ can desorb (with rate constant k_2) into the gas phase or proceed to the third step where it dissociates to give a Cu deposit and a gas phase by-product, $Cu(hfac)_2$.

Based on this reaction scheme, the Cu deposition rate, r_z , relates to the Langmuir - Hinshelwood rate equation and is expressed as

$$r_z = \frac{k_s P_A^2}{(K_d + P_A)^2} = \frac{P_A^2 k_{s0} \exp(-E_s/R_g T)}{(K_{d0} \exp(-\Delta H/R_g T) + P_A)^2}, \quad (3.2.41)$$

where E_s is the activation energy for the surface reaction, P_A is the precursor partial pressure, ΔH is the gain in energy resulting from direct desorption of the intermediate adsorbate in step 2 above, $Cu(hfac)$, and $K_d = \frac{k_2}{k_1}$.

Equations (3.2.38) and (3.2.39) were solved at steady state using the Galerkin finite element method. The resulting temperature distribution was used with Eqs. (3.2.40) and (3.2.41) to obtain the growth increment $\delta(x, y)$. The new deposit surface resulting from δ increment was then employed to determine the new surface temperature distribution from which $\delta(x, y)$ for the next time step is determined. This iteration was continued to simulate the evolution of the temperature distribution and deposit shape. A steady state solution was obtained also by setting the time derivatives in Eqs. (3.2.38)-(3.2.40) to zero. From Eq. (3.2.40), it is seen that

$$\delta(x, y) = \frac{1}{v_x} \int_\infty^x r_z(T_s(x', y)) dx'. \quad (3.2.42)$$

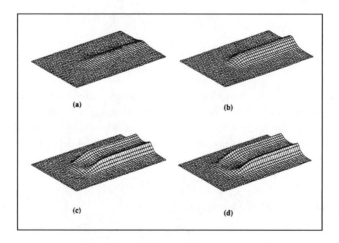

Fig. 3.2.9. Model simulation of copper deposits for different laser powers: (**a**) 0.46W, (**b**) 0.62W, (**c**) 0.94W, (**d**) 1.09W (reprinted from Han and Jensen 1994 with permission from the American Institute of Physics)

The volumetric growth rate G (cm^3/s) can be determined by multiplying the fully developed cross-sectional area of the deposit by the scan speed. This is expressed as

$$G = 2v_x \int_0^\infty \delta(-\infty, y) dy = 2 \int_0^\infty \int_\infty^{-\infty} r_z(T_s(x,y)) dx dy. \qquad (3.2.43)$$

The model was used to simulate the deposition of Cu stripes on a Si substrate for different laser powers ($0.46W$, $0.62W$, $0.94W$, and $1.09W$). The laser spot radius used was 4 μm, scan speed $10\mu m$, and precursor partial pressure 0.2 Torr. It is seen from Fig. (3.2.9) that so-called volcano-shaped deposit surfaces developed at higher laser powers which was in agreement with what was observed experimentally. This phenomenon was explained as being caused by increased desorption of the absorbed intermediate species, $Cu(hfac)$, caused by higher temperatures at the center of the beam.

Existing models on LCVD simulate the deposition of a microstructure (stripe, spot, or rod) given the employed parameters of the process. A more useful model is one that enables the determination of parameters to be employed in order to deposit or grow a microstructure with a pre-specified geometry. Such an approach is closely related to process control and optimization and is essential for rapid progress towards LCVD commercial applications.

A model to determine the dwell times during surface scanning in order to deposit a microstructure with pre-specified geometry has been developed by Nassar et al. (2002a). The heat equations in a three-dimensional LCVD that describe heat flow through the deposit and substrate are

$$c_D \rho_D \frac{\partial T_D}{\partial t} = \nabla(\kappa_D \nabla T_D) + Q_{in}, \qquad (3.2.44)$$

and

$$c_s \rho_s \frac{\partial T_s}{\partial t} = \nabla(\kappa s \nabla Ts), \qquad (3.2.45)$$

where T_D and T_s are the temperatures for deposit and substrate, respectively, and Q is the heat source from the laser beam. Also, c_D, ρ_d, κ_D, and c_s, $\rho_s \kappa_s$ are the specific heat, mass density, and thermal conductivity of the deposit and substrate, respectively. The interfacial equations for the deposit and substrate are

$$T_D = T_s \qquad \text{and} \qquad \kappa_D \frac{\partial T_D}{\partial z} = \kappa_s \frac{\partial T_s}{\partial z}. \qquad (3.2.46)$$

On the surface of deposit ($z = h_0(x,y)$),

$$-\kappa_D \frac{\partial T_D}{\partial z} = J_{loss},$$

and at the interface ($z = 0$)

$$\kappa_s \frac{\partial T_s}{\partial z} = J_{loss}.$$

During a time interval Δt, the local growth at a point of deposition may be defined by the Arrhenius equation as

$$\Delta \mathbf{z}_\eta(x,y) = k_0 \exp(-\frac{E_a}{R_g T_D})\hat{\eta}(x,y)\Delta t. \tag{3.2.47}$$

The unit normal to the surface of deposit, $z = h_0(x,y)$, is

$$\hat{\eta}(x,y) = \frac{1}{\varphi}\left[-\frac{\partial h_0}{\partial x}\hat{x} - \frac{\partial h_0}{\partial y}\hat{y} + \hat{z}\right], \quad \varphi = \sqrt{1 + (\frac{\partial h_0}{\partial x})^2 + (\frac{\partial h_0}{\partial y})^2}, \tag{3.2.48}$$

where $\hat{x}, \hat{y}$ and $\hat{z}$ are the unit vectors on the x, y, and z coordinates, respectively.

Equations (3.2.44), (3.2.45), and (3.2.47) may be solved numerically to obtain the dwell times during scanning for depositing a three-dimensional microstructure with pre-specified geometry. The deposition takes place layer by layer as shown schematically in Fig. (3.2.10). Let $z = h_0(x,y)$ represent the surface of the first layer and $z = h_1(x,y)$ that of the second layer. One needs to determine the dwell time at each point on the surface $z = h_0(x,y)$ needed for depositing the next layer or surface $z = h_1(x,y)$. We approximate the two surfaces by second degree polynomials as follows:

$$h_0(x,y) = a_0 x^2 + b_0 xy + c_0 y^2 + d_0 x + e_0 y + f_0, \tag{3.2.49}$$

and

$$h_1(x,y) = a_1 x^2 + b_1 xy + c_1 y^2 + d_1 x + e_1 y + f_1. \tag{3.2.50}$$

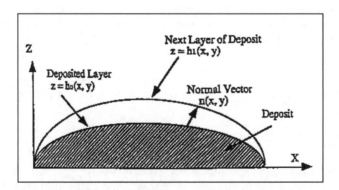

Fig. 3.2.10. Cross section of two consecutive layers of a parabolic surface (reprinted from Nassar et al. 2002a with permission from Elsevier Science)

In order to determine the growth $\Delta z_\eta(x_0, y_0)$ in Eq. (3.2.47) along a normal line between a point (x_0, y_0, z_0) on the surface $z = h_0(x,y)$ and the next surface

$z = h_1(x, y)$, one needs to find the normal line at (x_0, y_0, z_0) and its intersect point (x_1, y_1, z_1) on the surface $z = h_1(x, y)$. It is known from calculus that the equation for the normal line is given by

$$x = x_0 - \frac{\partial h_0}{\partial x}\rho, \ y = y_0 - \frac{\partial h_0}{\partial y}\rho, \ z = z_0 + \rho, \qquad (3.2.51)$$

where ρ is a parameter to be determined. Substituting Eq. (3.2.51) into Eq. (3.2.50) and solving for ρ, one can determine the intersect point (x_1, y_1, z_1) on the surface $z = h_1(x, y)$. Hence,

$$|\Delta \mathbf{z}_\eta(x_0, y_0)| = \sqrt{(x_1 - x_0)^2 + (y_1 - y)^2 + (z - z_0)^2} = \varphi |\rho|, \qquad (3.2.52)$$

where φ is defined in Eq. (3.2.48).

To determine the dwell times, we decompose the surface into a number of small subelements, Δ_{ij}, $i = 1, \cdots, M$, and $j = 1, \cdots, N$. The temperature is uniform in each subelement. Let the temperature in subelement Δ_{pq} be $(T_D)_{ij}(x_p, y_q, h_0(x_p, y_q))$ when the laser beam is focused in subelement Δ_{ij}. We assume that the temperature is at steady state. From Eqs. (3.2.47) and (3.2.52), the laser beam dwell time Δt_{ij} in each subelement Δ_{ij} can be calculated from the linear system

$$\sum_{i=1}^{M}\sum_{j=1}^{N} k_0 \exp\left[-\frac{E_a}{R_g (T_D)_{ij}(x_p, y_q, h_0(x_p, y_q))}\right] \Delta t_{ij} = |\Delta \mathbf{z}_\eta(x_0, y_0)|.$$
$$(3.2.53)$$

The temperature distribution on the surface, $(T_D)_{ij}(x_p, y_q, h_0(x_p, y_q))$, can be determined by solving Eqs. (3.2.44) and (3.2.45) in the steady state condition. However, since one is dealing with a relatively thin deposit, one can simplify the equations in a way similar to that used by Arnold and Baeuerle (1993b). Integrating Eq. (3.2.44) in the steady state between zero and $h = h_0(x, y)$, one obtains

$$-\int_0^{h_0}\left[\frac{\partial}{\partial x}(\kappa_D \frac{\partial T_D}{\partial x}) + \frac{\partial}{\partial y}(\kappa_D \frac{\partial T_D}{\partial y})\right] dz - \int_0^{h_0}\frac{\partial}{\partial z}(\kappa_D \frac{\partial T_D}{\partial z})dz$$
$$= \int_0^{h_0} Q_{in}dz. \qquad (3.2.54)$$

Assume that $T_D(x, y, z) \approx T_D(x, y, 0)$, which holds if $\frac{h_0}{r_D \kappa^*} \ll 1$, where $\kappa^* = \frac{\kappa_D}{\kappa_s}$ and r_D is the radius of the base of deposit. With this approximation, Eq. (3.2.54) becomes

$$-\frac{\partial}{\partial x}(h_0 \kappa_D \frac{\partial T_D}{\partial x}) - \frac{\partial}{\partial y}(h_0 \kappa_D \frac{\partial T_D}{\partial y})$$
$$= \int_0^{h_0} Q_{in}dz + J_{loss}(z = h_0) - J_{loss}(z = 0). \qquad (3.2.55)$$

Since the thermal conductivity of the gas is much smaller than that of the solid deposit, heat loss from the surface to the gas phase may be ignored (*i.e*, $J_{loss}(z = h_0) = 0$), and we let $J_{loss}(z = 0) = \kappa_s \frac{\partial T_s}{\partial z}$ at the interface. Also, we assume that the laser energy which is Gaussian is totally absorbed within the deposit which gives

$$\int_0^{h_0} Q_{in} dz = I_{ij}(x, y)$$

$$= \frac{P_0(1 - \Gamma(\phi))}{2\pi\sigma^2}$$

$$\cdot \exp\left[-\frac{(x - x_i)^2 + (y - y_i)^2}{2\sigma^2}\right] \langle \hat{z}, \hat{\eta} \rangle_{(x,y)}, \quad (3.2.56)$$

where P_0 is the laser power, σ is the standard deviation of the Gaussian distribution of the laser beam, and $\langle \hat{z}, \hat{\eta} \rangle$ is the inner product of $\hat{z}$ and $\hat{\eta}$. The angular spectral reflectance $\Gamma(\phi)$ is defined as (Seidel and Howell, 1992)

$$\Gamma(\phi) = \frac{1}{2}\left\{\left[\frac{\cos\phi - \sqrt{\eta_i^2 - \sin^2\phi}}{\cos\phi + \sqrt{\eta_i^2 - \sin^2\phi}}\right]^2 + \frac{\eta_i^2\cos\phi - \sqrt{\eta_i^2 - \sin^2\phi}}{\eta_i^2\cos\phi + \sqrt{\eta_i^2 - \sin^2\phi}}\right\},$$

where η_i is the deposit index of reflection and ϕ is the local incidence angle. Equation (3.2.55) reduces now to

$$\kappa_s \frac{\partial T_s}{\partial z} = I_{ij} + \frac{\partial}{\partial x}(h_0\kappa_D\frac{\partial T_D}{\partial x}) + \frac{\partial}{\partial y}(h_0\kappa_D\frac{\partial T_D}{\partial y}). \quad (3.2.57)$$

Eq. (3.2.57) can be considered as a modified boundary condition for calculating the temperature in the substrate. Equation (3.2.45) in steady state can now be expressed as

$$\frac{\partial}{\partial x}(\kappa_s \frac{\partial T_s}{\partial x}) + \frac{\partial}{\partial y}(\kappa_s \frac{\partial T_s}{\partial y}) + \frac{\partial}{\partial z}(\kappa_s \frac{\partial T_s}{\partial z}) = 0. \quad (3.2.58)$$

To obtain the dwell time at each point on the surface $z = h_0(x, y)$ in order to obtain the next deposited layer $z = h_1(x, y)$ one may solve numerically for the temperature distribution $(T_s)_{ij}(x, y, z)$ from Eq. (3.2.58) with Eq. (3.2.57) as boundary condition. On other surfaces, the temperature is set equal to the surrounding temperature. Let $(T_D)_{ij}(x, y, z) \approx (T_D)_{ij}(x, y, 0) \approx (T_s)_{ij}(x, y, 0)$, and solve the linear system in Eq. (3.2.53) to obtain the dwell times Δt_{ij} ($i = 1, \cdots M; J = 1, \cdots N$). Repeating this iterative process layer by layer, one obtains the desired microstructure.

The model is used to simulate the deposition of a microlens from nickel on a graphite substrate. The radius of the lens is 10 μm, and its height at the center is 2.5 μm. Parameters employed were $E_a = 9.4 \times 10^4 J/mol$, $R_g = 8.314$ $J/mol \cdot K$, $k_0 = 1.37 \times 10^4 mm/s$, $\kappa_D = 6.55 \times 10^{-2} W/mm \cdot K$, $\kappa_s = 1.7 \times 10^{-3}$

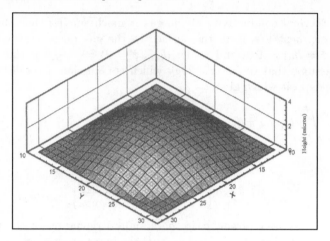

Fig. 3.2.11. Model simulation of a parabolic microlens (reprinted from Nassar et al. 2002a with permission from Elsevier Science)

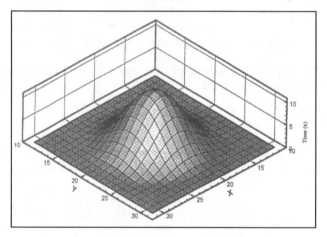

Fig. 3.2.12. Total dwell times for each subelement on the surface that correspond to the microlens deposit in Fig. (3.2.11) (reprinted from Nassar et al. 2002a with permission from Elsevier Science)

$W/mm \cdot K$, $T_\infty = 475K$, $P_0 = 0.17W$, and $\sigma = 0.0005$ mm. The index of reflection was $\eta_i = 3.95$. Figure (3.2.11) shows the shape and height of the microlens for 20 layers of deposit. Figure (3.2.12) gives the corresponding dwell times. As expected, dwell times agree with the deposit shape. Figure (3.2.13) presents the change in maximum and minimum surface temperatures as a function of the height of the microlens or number of layers of deposit. It is seen that surface temperature decreases with an increase in the height of the deposit. The model enables one to adjust the laser power in order to prevent the surface temperature from falling and growth from slowing down or coming to a halt. In this simulation, thermal conductivities were assumed to be independent of temperature. Equations. (3.2.57) and (3.2.58) can be solved numerically for the case where conductivities are temperature dependent. When thermal conductivity increases with temperature, one expects a lower temperature at the surface than would be the case for a constant conductivity. This is caused by enhanced conductivity away from the surface with increasing temperatures.

A model based on the axisymmetric rod growth heat equation in Dai et al. (1999) was developed for optimizing the laser power required for inducing the growth of an axisymmetric rod of a pre-specified geometry (Nassar et al 2002b). For an example, given the specified geometry of the rod to be cylindrical, one may determine the expected growth, $\mathbf{R}_n$, based on this geometry and obtain, using the Arrhenius relation in Eq. (3.2.25), the temperature at the surface of the rod that corresponds to the expected growth. From Eqs.(3.2.31)-(3.2.37) coupled with Eqs.(3.2.26)-(3.2.28) one may calculate the temperature distribution on the surface of the rod that corresponds to a certain laser power. To optimize the power, one may solve for the power (using least squares techniques) that minimizes the sum of squared deviations between expected temperatures (from the Arrhenius equation) and calculated temperatures based on the heat equation. Figure (3.2.14) demonstrates the application of the numerical model. The first step is to calculate the expected growth, $\mathbf{R}(x, r_n(x))$, from the pre-specify geometry of the current surface, $\mathbf{R}(x, r(x))$.

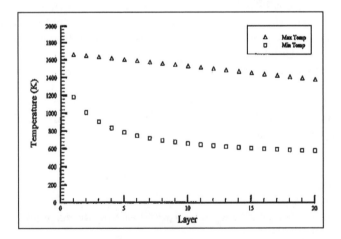

Fig. 3.2.13. Maximum and minimum surface temperatures as a function of deposit height in number of layers (reprinted from Nassar et al. 2002a with permission of Elsevier Science)

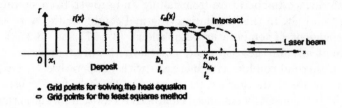

Fig. 3.2.14. Schematic diagram of growth of a cylindrical rod (reprinted from Nassar etal. 2002b with permission of Allerton Press Inc., New York)

It is seen, from Fig.(3.2.14), that

$$r(x) = v\sqrt{l_2 + c} - l_1, \qquad 0 \le x \le l_1$$

$$= v\sqrt{l_2 + c - x}, \qquad l_1 \le x \le l_2, \tag{3.2.59}$$

and

$$r_n(x) = v\sqrt{l_2 + c} - l_1, \qquad 0 \le x \le l_1 + a$$

$$= v\sqrt{l_2 + c + a - x}, \qquad l_1 + a \le x \le l_2 + a, \tag{3.2.60}$$

where v, c, l_1, l_2 are constants and a is the expected growth at the tip. For $l_1 \le x \le l_2$, the surface of the rod is a portion of a parabola which is used to simulate the shape of the rod near the tip. To simulate the growth process, one calculates for each grid point on the rod surface, $r(x)$, between l_1 and l_2 the normal line (using the slope-point equation of a straight line) and determines the intersect of this normal line on the expected growth curve, $r_n(x)$. The next step is to calculate for each grid point the distance along the normal line between the two surfaces, $r(x)$ and $r_n(x)$. Using this distance in the Arrhenius equation, one calculates the expected temperature $(T_d)_{bi}^e$. From Eq.(3.2.31), one may calculate the temperature, $(T_d)_{bi}(P_0)$, for a given laser power, P_0. An optimum solution for P_0 for each layer of growth can be obtained by minimizing the expression

$$S(P_0) = \sum_{i=1}^{N_g} [(T_d)_{bi}^e - (T_d)_{bi}(P_0)]^2. \tag{3.2.61}$$

Differentiating $S(P_0)$ with regard to P_0 and setting the results equal to zero gives

$$\frac{\partial S(P_0)}{\partial P_0} = \sum_{i=1}^{N_g} \frac{\partial (T_d)_{bi}(P_0)}{\partial P_0} [(T_d)_{bi}^e - (T_d)_{bi}(P_0)] = 0. \tag{3.2.62}$$

In matrix form, Eq.(3.2.62) may be expressed as

$$\mathbf{X}^T \cdot [\mathbf{T}_d^e - \mathbf{T}_d(P_0)] = 0, \tag{3.2.63}$$

where

$$\mathbf{X}^T = \left[\frac{\partial (T_d)_{b_1}(P_0)}{\partial P_0}, ..., \frac{\partial (T_d)_{b_{N_g}}(P_0)}{\partial P_0} \right], \tag{3.2.64}$$

$$\mathbf{T}_d^e = \left[(T_d)_{b_1}^e, ..., (T_d)_{b_{N_g}}^e \right]^T, \tag{3.2.65}$$

and

$$\mathbf{T}_d = \left[(T_d)_{b_1}, ..., (T_d)_{b_{N_g}} \right]^T. \tag{3.2.66}$$

Equation(3.2.62) may be solved for P_0 (using the Newton iteration method) which is used to grow the new surface, $\mathbf{R}(x, r_n(x))$. This new surface is in turn reset at the current surface, $\mathbf{R}(x, r(x))$ and the process repeated layer by layer until the desired rod is obtained.

To demonstrate the applicability of the model, an experiment was carried out in which amorphous carbon (using methane (CH_4) as a precursor gas) was deposited onto a graphite substrate. Parameters used in the Arrhenius equation were $E_a = 1.82 \times 10^5$ (j/mol), $R_g = 8.314$ $(J/mol.K)$,and $k_0 = 2.37 \times 10^4$ mm/s. Parameters in Eqs.(3.2.31)-(3.2.37) were $k_d = 1.65$ $W/mm.K$, $k_s = 1.7 \times 10^{-3}$ $W/mm.K$, $T_\infty = 300K$, $\omega = .01mm$, and $A = 1$. The heat transfer coefficient h_{conv} was obtained from the relation (Maxwell 1966)

$$h_{conv} = \frac{N_u k_{gas}}{2r(x)}, \tag{3.2.67}$$

where $N_u =$Nusselt number $= 0.36$ and $k_{gas} = 0.001$.

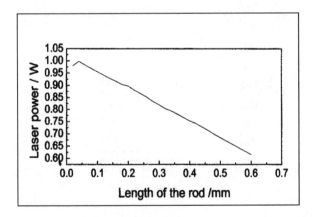

Fig. 3.2.15. Laser power calculated from the model as a function of rod length (reprinted from Nassar et al. 2002b with permission of Allerton Press Inc., New York)

Based on these parameters, laser power was calculated from the model for the purpose of growing a cylindrical rod with a diameter of 0.08 mm, tip (at $x = l_2$, in Eq.(3.2.59)) equal to 0.015 mm, and a length of 0.6 mm. Figure (3.2.15) shows a plot of the laser power. As expected, the power decreased with an increase in the height of the rod. Figure (3.2.16) presents the surface temperature for rods of different lengths. It is seen that the surface temperature increased from the base to the top of the rod. Also, the base temperature is higher for short than for long rods. Figure (3.2.17a) shows the shape of a rod grown by using a constant laser power of 0.6W. Figure (3.2.17b,c) show rods grown using the laser powers from the model as shown in Fig.(3.2.15). It is seen that rods grown based on the powers from the model are closer to being cylindrical in shape than the rod grown under a constant power. Rod diameter measurements are shown in Fig. (3.2.18) which confirm the shapes shown in Fig.(3.2.17). It is seen from the figure that the diameter of the deposited rod does not reach its expected value of 0.08 mm until it reaches a height of about 125 μm. This relatively small discrepancy between observed and expected diameter values during early growth may be due to the temperature not reaching steady state as assumed in the model. This assertion can be tested by using the unsteady state heat equation in the model. Another point to investigate is the initial condition with regard to the substrate temperature. Preliminary investigations with a nickel deposit showed that heating the substrate surface initially to above ambient reduced the discrepancy between observed and expected diameter values.

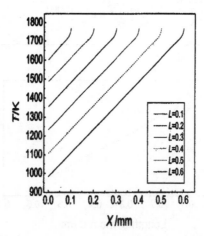

Fig. 3.2.16. Surface temperature as a function of different rod lengths (reprinted from Nassar et al. 2002b with permission of Allerton Press Inc., New York)

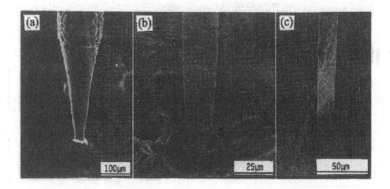

Fig. 3.2.17. Rod deposits (**a**) without model optimization and (**b**) with model optimization (reprinted from Nassar et al. 2002b with permission of Allerton Press Inc., New York)

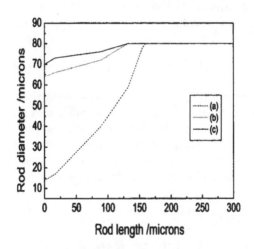

Fig. 3.2.18. Rod diameter, for the rods in Fig. (3.2.17), as a function of rod length (reprinted from Nassar et al. 2002b with permission of Allerton Press Inc., New York)

3.3 Mass Transport Limited Deposition

In an LCVD process, the deposition is initially kinetically limited. With increasing surface temperature, the reaction rate increases exponentially according to the Arrhenius law. This increase in reaction rate causes a relatively large conversion of reactants which eventually leads to a deficiency in the precursor at the surface of the reaction zone. This deficiency causes the reaction rate to slow and the deposition rate to level off. At this stage, the process becomes diffusion-limited, and the deposition rate is dependent on the gas flux at the surface. If the surface is relatively large as in a scanning operation, it is possible, depending on the temperature, to have some portions of the deposit growing at the flux rate while other portions at a kinetically limited rate. In such a case, diffusion-limited and kinetically limited deposition rates may be applied to different portions of the deposit, depending on the surface temperature, to simulate growth.

A model was presented by Skouby and Jensen (1988) which considers heat conduction through the solid-phase, heat transfer through the gas-phase, and gas mass transport through diffusion to the surface of deposit. The transient heat conduction equation for characterizing the temperature distribution in the solid is given by

$$\rho C_p \frac{\partial \theta}{\partial t} = \nabla \left(k(\theta) \nabla \theta \right) + \frac{Q}{T_0}. \tag{3.3.1}$$

Here, θ is a scaled temperature and is equal to $\frac{T-T_0}{T_0}$, where T_0 is the initial temperature. The density of the solid is ρ, C_p is the heat capacity of the solid, and $k(\theta)$ is the thermal conductivity. The heat source term for a Gaussian beam is expressed as

$$Q(r, z) = [\alpha P(1 - \Re)/\pi \omega^2] \exp(-az) \exp(-r^2/\omega^2), \tag{3.3.2}$$

where P is the laser power, α is the absorption coefficient of the substrate, $\Re$ is the surface reflectivity, r and z are the radial and axial dimensions in cylindrical coordinates, and ω is the Gaussian beam width which is defined as the radius at which the intensity of the beam is $1/e$ times its maximum. The boundary conditions at the center line ($z = 0, r = 0$) assume symmetry and no energy flux from the solid surface to the gas phase. These are expressed as

$$\frac{\partial \theta}{\partial R}\Big|_{R=0} = 0, \quad \text{and} \quad \frac{\partial \theta}{\partial Z}\Big|_{Z=0} = 0. \tag{3.3.3}$$

Here, $R = r/\omega$ and $Z = z/\omega$. For large R and Z, Robin boundary conditions are imposed giving the expressions

$$\frac{\partial \theta}{\partial R} = \frac{-\theta R}{R^2 + Z^2} \quad \text{and} \quad \frac{\partial \theta}{\partial Z} = \frac{-\theta Z}{R^2 + Z^2}. \tag{3.3.4}$$

In deriving the Robin boundary conditions, the flux across a spherical boundary or shell far from the origin is assumed to be constant.

The deposition system considered was nickel (from nickel carbonyl as the precursor gas) deposited on a semi-infinite quartz substrate. As such, the empirical expression used for the temperature-dependent thermal conductivity in W/mK was $k(T) = \exp[a + b \ln T + c(\ln T^2)]$, where $a = 29.23$, $b = -9.84$, and $c = 0.837$. This expression predicts an increase in thermal conductivity with an increase in temperature.

The surface reflectivity was given by the empirical formula $\Re = \Re(1) - [\Re(1) - \Re(2)]\exp(-\gamma l)$, where $\Re(1) = 0.94$ for nickel, $\Re(2) = 0.12$ for the SiO_2 substrate, l is the thickness of the nickel deposit, and γ is a parameter determined empirically.

In the numerical solution to Eq. (3.3.1), the substrate alone, without the nickel deposit, was included in the solid phase. The deposit, therefore, acted as a filter for the substrate by attenuating the beam intensity. This procedure was adopted in order to avoid having a stiff numerical solution caused by the small time steps that would be required had the thin deposit been included in the solid phase.

The gas is heated by heat conduction from the surface of the substrate. The process is considered to be in steady state. The heat conduction equation in the gas-phase is then

$$\nabla[k(\theta)\nabla\theta] = 0, \tag{3.3.5}$$

where $k(\theta)$ is the thermal conductivity and θ is the normalized temperature rise $\frac{T-T_0}{T_0}$ of the gas. The time-dependent thermal conductivity was assumed to be a function of the square root of temperature. A value of $1.0 \times 10^{-2}\ W/mK$ was used for $k(\theta)$ at $300K$. The boundary conditions were the same as those in Eqs. (3.3.3) and (3.3.4). The boundary condition at the interface between solid and gas was

$$\theta \mid_{Z=0} = \theta_s,$$

where θ_s is the normalized surface temperature rise calculated in the solid-phase from Eq. (3.3.1). The normalized temperature rise in the gas phase, θ_g, obtained from the solution to Eq. (3.3.5) is required for the gas phase mass transfer calculations.

In general, the thermal dissociation reaction of the gas molecules (AB_v) that diffuse to the heated surface of the solid can be expressed as

$$AB_v\ (g) \rightarrow A_s + vB_g$$

Here, an AB_v gas molecule reacts to form a solid A_s molecule and v molecules of gas B that diffuse away from the surface. In the case of Nickel carbonyl, the reaction is

$$Ni(CO)_4(g) \rightarrow Ni(s) + 4CO(g).$$

The molar flux of the gas $(Ni(CO)_4)$ in the z (N_Z) and r (N_r) directions can be expressed by the equations

$$N_z = \left[-cD(\frac{\partial x}{\partial z} + a_T x(1-x)\frac{\partial \ln T}{\partial z})\right]/[1 + x(v-1)], \tag{3.3.6}$$

$$N_r = \left[-cD(\frac{\partial x}{\partial r} + a_T x(1-x)\frac{\partial \ln T}{\partial r})\right]/[1 + x(v-1)], \tag{3.3.7}$$

where a_T is the thermal diffusion coefficient, x is the mole fraction of gas AB_v, c is the total molar concentration, D is the binary diffusion coefficient for the gas system, and T is the gas-phase temperature.

A symmetry boundary condition at the center line is given by the equation

$$\frac{\partial x}{\partial r}\Big|_{r=0}= 0.$$

The boundary conditions far from the origin (large r and z values) are

$$\frac{\partial x}{\partial r} = -\frac{[1 + x(v - 1)]r}{(r^2 + z^2)(v - 1)} \times \ln\left(\frac{1 + x(v - 1)}{1 + x_b(v - 1)}\right), \qquad (3.3.8)$$

$$\frac{\partial x}{\partial z} = -\frac{[1 + x(v - 1)]z}{(r^2 + z^2)(v - 1)} \times \ln\left(\frac{1 + x(v - 1)}{1 + x_b(v - 1)}\right), \qquad (3.3.9)$$

where x_b is the mole fraction far from the laser beam and may be assumed to be 1. In the case $v = 1$, these boundary conditions reduce to

$$\frac{\partial x}{\partial r} = \frac{(x_b - x)r}{r^2 + z^2} \quad \text{and} \quad \frac{\partial x}{\partial z} = \frac{(x_b - x)z}{r^2 + z^2}. \qquad (3.3.10)$$

The boundary condition at the solid surface is obtained by equating the flux in the z direction to the reaction rate at the surface, r_s. This gives

$$\frac{-cD(\frac{\partial x}{\partial z})}{1 + x(v - 1)} = r_s(p, T_s, x). \qquad (3.3.11)$$

In their calculations, the authors used two expressions,

$$r_s(p, T_s, x) = k_0 \exp(-E_a/RT)px, \qquad (3.3.12)$$

(where k_0 is a constant in Mol per Torr per unit area per *sec*, E_a is the activation energy, R is the ideal gas constant, T is temperature, and p is the total pressure) and the Langmuir-Hinshelwood equation

$$r_s(p, T_s, x) = \frac{k_s \exp(E_s/RT)px}{1 + K_D \exp(E_D/RT)px}, \qquad (3.3.13)$$

where k_s, E_s, K_D and E_D are Langmuir-Hinshelwood parameters. From the above mass transport equations, the concentration x is calculated, and the surface reaction rate, r_s is determined. From this reaction rate, the deposit is allowed to grow for a time period equal to the time step for the calculation of the temperature distribution from Eq. (3.3.1). Since the temperature in the solid changes as the deposit grows in thickness, the parameters in Eq. (3.3.1) must be altered and the equation solved for another time step. This gives rise to another period (equal to the time step) of film growth. This iteration is repeated until a certain film thickness or growth time is reached.

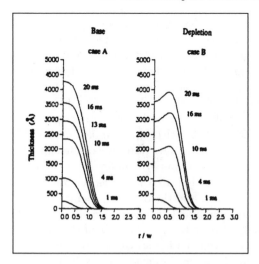

Fig. 3.3.1. Deposit thickness profile at different growth times: (**A**) base case, (**B**) depletion case (reprinted from Skouby and Jensen 1988 with permission from the American Institute of Physics)

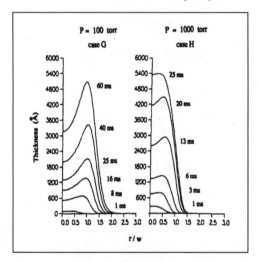

Fig. 3.3.2. Deposit thickness profile at different times of growth using the Langmuir-Hinshelwood kinetics (reprinted from Skouby and Jensen 1988 with permission from the American Institute of Physics)

Numerical calculations were accomplished using the Galerkin finite element method. Results showed that the radial temperature profiles in the solid were Gaussian-like with the maximum temperature occurring at the center of the deposit spot as would be expected because of the Gaussian beam intensity.

The mole fraction, x, near the center of the laser beam varied between 0.01 and 1.0, depending on the relative magnitudes of growth and diffusivity. The mole fraction, x, far from the center of the beam approached 1.0 as expected. The growth of volcano-like deposits could be explained from the model based on depletion effects or nonlinear kinetic effects. Figure (3.3.1) demonstrates volcano-like growth based on the depletion effect. Figure (3.3.1 A) shows a deposition profile flat around the laser beam center. For this profile, the reaction rate was that in Eq. (3.3.12) with $k_0 = 1 mol/Torr/cm^2/\sec$, $E_A = 15000$ $cal/mole$, and $p = 100\ Torr$. The diffusion coefficient $D = 1 cm^2/s$, $v = 1$, and $a_T = 0$. The laser power was $3W$, and γ for reflectivity was $6 \times 10^6 m^{-1}$. The volcano-like deposit in Fig. (3.3.1 B) was generated by changing the pressure from 100 to 500 $Torr$, and the diffusion coefficient to $D = 0.2 cm^2/s$. The volcano-like deposit is explained by a depletion effect caused by a larger growth rate relative to the diffusion rate. In support of this depletion hypothesis, the mole fraction, x, was about 0.5 in case A, whereas it approached 0.01 in case B. Volcano-like deposits were obtained also by increasing k_0 without changing the pressure or the diffusion coefficient. Figure (3.3.2) demonstrates the effect of nonlinear reaction kinetics, (Eq. 3.3.13) on the formation of volcano-like deposits. The volcano-like deposit in Fig. (3.3.2 G) was generated using the parameters $k_s = 2 \times 10^{-7}\ mol/Torr/cm^2/s$, $K_D = 1 \times 10^{-4}\ Torr^{-1}$, $E_s = 3400\ cal/mol$, and $E_D = 16000\ cal/mol$. The values for the other parameters were $v = 1$, $a_T = 0$, $\gamma = 6 \times 10^6\ m^{-1}$, $p = 100 Torr$, and $D = 1 cm^{-2}/s$. The parameters in Fig. (3.3.2 H) were the same as those in case G, except that the pressure was changed to 1000 $Torr$ and the diffusion coefficient D to $0.1\ cm^2/s$. The relative reduction in volcanoes was explained as being caused perhaps by the fact that the changes in pressure and diffusivity raised the temperature for the maximum reaction rate from Eq. (3.3.13) above the maximum solid surface temperature. This phenomenon may have reduced the reaction rate relative to the diffusion rate and prevented the depletion effect from occurring. It is likely, however, that the volcanoes in case G are not caused by a depletion effect, but rather by the nonlinear reaction rate. Thus, nonlinear kinetics may account for experimentally observed volcano-like deposits in many organometallic compounds.

Basically the same model was used by Yankova and Coply (1993) to simulate the deposition of aluminum from trimethylaluminum, $Al(CH_3)_3$, gas in the presence of argon. In their treatment, they considered the gas phase heat equation under a transient condition and a carrier gas in the gas phase mass transfer. Volcano-like deposits were predicted for high laser power densities.

The effects of temperature dependencies in particle density, molecular diffusion coefficient, and thermal conductivity on the reaction kinetics in laser-induced chemical vapor deposition have been investigated by Baeuerle et al. (1990). The reaction zone is assumed to be a hemisphere of radius r_D on a semi-infinite substrate with spherical symmetry relative to the zone center. This experimental situation applies to deposition of spots where the laser beam is Gaussian. The surface temperature of the deposit is assumed uniform and equal to the temperature of the ambient, $T(r_D)$, at the surface, $r = r_D$. The reaction of the gas AB_v at the surface is of the form

$$AB_v + G \rightarrow A_s + B_g + G, \tag{3.3.14}$$

where G is a carrier gas. The decomposition of AB_v is assumed to be first order. Also, in what follows, equimolecular reactions where $v = 1$ are assumed which implies that the total number of molecules remains constant.

Under stationary conditions, the reaction rate is independent of time and may be expressed as

$$W = kN_{AB}(r_D) = kN(r_D)x_{AB}(r_D), \tag{3.3.15}$$

where $N_{AB}(r_D)$ is the particle density of gas AB at $r = r_D$, $N(r_D)$ is the total particle density, and $x_{AB}(r_D)$ is the molar fraction of gas AB. The rate constant k according to the Arrhenius law is given by $k = k_0 \exp(-E_a/RT_s)$. The particle density $N_i(r)$ is assumed to change only with the temperature distribution, T_r, in the ambient. The total particle density, $N(r)$, is thus the sum of $N_{AB}(r)$, $N_B(r)$, and $N_G(r)$. It is assumed that

$$N(r) = N(\infty) \left(\frac{T(\infty)}{T(r)} \right)^q, \tag{3.3.16}$$

where $q = 1$ for an ideal gas. Heating of the ambient is assumed to arise through energy flux from the laser-heated deposit surface ($r = r_D$) into the gas phase. Heat transport is assumed to occur through heat diffusion only.

In order to determine the effects of temperature dependencies in material properties, the equation of continuity for the gas species AB and of heat for the temperature in the gas phase must be considered jointly. The continuity equation is given by

$$-\nabla j_{AB}(r) = \frac{1}{r^2} \nabla[r^2 N(r) D_{AB}(r) \nabla x_{AB}(r)] = 0, \tag{3.3.17}$$

where D_{AB} is the diffusion coefficient of species AB in the gas mixture, and j_{AB} is the flux of species AB. The boundary conditions for Eq. (3.3.17) are

$$x_{AB}(r \rightarrow \infty) = x_{AB}(\infty) \qquad \text{and}$$

$$-j_{AB}(r_D) = N(r_D)D_{AB}(r_D) \left(\frac{\partial x_{AB}(r)}{\partial r} \right)_{r=r_D} = W \tag{3.3.18}$$

$$= kN(r_D)x_{AB}(r_D).$$

The heat equation for the temperature distribution in the ambient is

$$\nabla[r^2 \kappa(r) \nabla T(r)] = 0, \tag{3.3.19}$$

where $\kappa(r)$ is the thermal conductivity in the ambient medium. The boundary conditions are

$$T(r = r_D) = T_s \text{ and } T(r \to \infty) = T(\infty). \tag{3.3.20}$$

A solution to the mole fraction x_{AB} is given as

$$x_{AB}(r) = x_{AB}(\infty)(1 - \frac{F(r)}{\zeta + F(r_D)}), \tag{3.3.21}$$

where $\zeta = \frac{D_{AB}(r_D)}{k r_D}$, and $F(r) = r_D N(r_D) D_{AB}(r_D) \int_r^\infty \frac{dy}{y^2 N(y) D_{AB}(y)}$.

The solution to Eq. (3.3.19) gives the temperature distribution in the ambient.

$$T(r) = T(\infty) + \Delta T_s \frac{\int_r^\infty \frac{dy}{y^2 \kappa(y)}}{\int_{r_D}^\infty \frac{dy}{y^2 \kappa(y)}}, \tag{3.3.22}$$

where ΔT_s is the laser-induced temperature rise at the deposit surface $r = r_D$. The temperature-dependent diffusion coefficient and thermal conductivity were given in the form

$$D_{AB}(r) = D_{AB}(\infty) \left(\frac{T_g(r)}{T_{g(\infty)}} \right)^n, \tag{3.3.23}$$

and

$$\kappa(r) = \kappa(\infty) \left(\frac{T_g(r)}{T_{g(\infty)}} \right)^m, \tag{3.3.24}$$

where $T_g(r)$ is the temperature in the ambient or gas phase. Investigations of the model for arbitrary values of q, n, and m in Eqs. (3.3.16), (3.3.23), and (3.3.24) were performed. An important finding was that temperature-dependent diffusion increased the reaction rates in the mass transport limited region by approximately a factor of 10. On the other hand, temperature-dependent thermal conductivity had a minor effect on reaction rates. Further investigations of the model revealed that concentration-dependent thermal conductivity had no significant effect on reaction rates. However, thermal diffusion can significantly reduce or increase reaction rates (in mass-transport and kinetically limited regimes) depending on the relative mass size of the precursor gas (AB) to that of the carrier gas (G). For $m_G > m_{AB}$, the reaction rate at the surface increases and it decreases if $m_G < m_{AB}$.

In a further study, Kirichenko et al. (1990) used the same modelling approach in Baeuerle et al. (1990) to investigate the case of non-equimolecular reactions where $v \neq 1$ in Eq. (3.3.14). Of interest is the effect of counter-diffusion of the reaction product in the ambient on the reaction rate at the solid surface. In this case, the continuity equation for species i ($i = AB_v$, B , G) under steady state conditions can be expressed in general as

$$\nabla J_i(x_\alpha) = 0, \tag{3.3.25}$$

where $J_i(x_\alpha)$ is the total flux of species i at the reaction zone surface coordinate x_α. Flux for species AB_v can be written as

$$J_{AB} = -ND_{AB}(\nabla x_{AB} + k_T \nabla \ln T_g) + \mu.N_{AB}, \tag{3.3.26}$$

where k_T is thermal diffusion and μ is the average particle velocity of the gas. A similar equation can be written for species B. Here, $k_T = \alpha_T \, x_{AB}(1-x_{AB})$ when $N_G = 0$ and $k_T = \alpha_T \, x_G$ when $N_G \gg N_{AB} + N_B$.

A continuity equation for the total particle density can be written in the form

$$\nabla[N(x_\alpha) \cdot \mu(x_\alpha)] = 0. \tag{3.3.27}$$

The heat equation is given in the from

$$c_\rho \nabla(NT_g\mu) - \nabla(\kappa \nabla T_g) = 0, \tag{3.3.28}$$

where c_ρ is the heat capacity.

The boundary conditions for Eq. (3.3.28) are those given in Eq. (3.3.20), and those for Eq. (3.3.25) in the case of AB are given in Eq. (3.3.18). For B, the boundary condition at the surface of the reaction zone is

$$J_B(r_s) = vW = vk(r_s)N_{AB}(r_s). \tag{3.3.29}$$

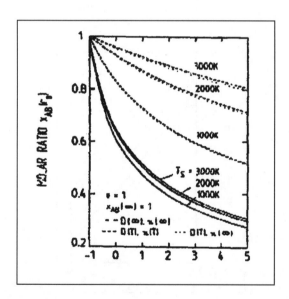

Fig. 3.3.3. Molar ratio x_{AB} at the surface of deposit as a function of b for different surface temperatures, T_S. Solid curves represent temperature independent parameters $(D(\infty), k(\infty))$, dotted curves represent temperature-dependent $D(D(T), k(\infty))$, and dashed curves represent temeperature-dependent D and $k(D(T),k(T))$. $\varepsilon = \dfrac{kr_D}{D_{AB}(\infty)} = 1$ and $x_{AB}(\infty)=1$ (reprinted from Kirichenko et al. 1990, Fig. 6, with permission from Springer-Verlag)

Various solutions of Eqs. (3.3.25) - (3.3.28) were investigated for a spherical reaction zone and under different scenarios. These include reactions with and without a carrier gas under an isothermal gas phase condition, and gas phase heating through heat flux from the reaction zone surface with temperature-independent and temperature-dependent parameters. The salient finding was that in the absence of a carrier gas, counter-diffusion effect of the reaction product in the gas phase can significantly reduce the surface reaction rate in the mass transport limited region. This reduction is most significant for v in the range -1 to 2 and for high concentrations of the mole fraction, $x_{AB}(\infty)$. There seems to be no counter-diffusion effects on reaction rates when a carrier gas is present in the ambient. Figure (3.3.3) demonstrates the effect of b $(b = v - 1)$ on reducing the molar ratio, x_{AB}, at the surface of the reaction zone for the case of gas phase heating with no thermal diffusion and no carrier gas.

Further investigations of the same model by Lukyanchuk et al. (1992) revealed that an increase in reaction rate at the surface could arise when ordinary and thermal diffusion are considered simultaneously. The thermal diffusion k_T is given by

$$k_T = \alpha_T x_{AB} x_G = \alpha_T^* x_{AB}, \qquad (3.3.30)$$

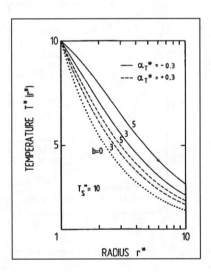

Fig. 3.3.4. Temperature distribution in the gas-phase as a function of r^* for surface temperature, $T_s^*=10$, thermal diffusion constant, $\alpha_T^*=-0.3$ (solid curves), $\alpha_T^*=-0.3$ (dashed curves), and different b values. For b=0 (dotted curve) the curve is independent of α_T^* (reprinted from Luk'yanchuk et al. 1992 with permission from Elsevier Science)

where α_T (the thermal diffusion constant) depends on the masses, m_i, of the different species in the gas mixture, $\alpha_T = \alpha_0(m_{AB} - m_G)/(m_{AB} + m_G)$; $\alpha_0 > 0$.

Figure (3.3.4) shows the effect of $b = v - 1$ and the thermal diffusion constant, α_T^*, on the gas temperature distribution in the ambient ($r^* = r/r_D$). In this calculation, the parameter values used were applicable to the deposition of carbon from hydrocarbons of the type $C_x H_y$. Carrier gases considered were H_2, He, and Ar. It is seen from the figure that the gas phase temperature out from the surface ($r^* = \frac{r}{r_D} = 1$) of the reaction zone increases as b increases. Also, this temperature increase is more pronounced when α_T^* is negative. This was attributed to a coupling effect between thermal diffusion and the so-called chemical convection (caused by the change in particle number density because of b being different from zero). The normalized molar ratio $x_{AB}^*(1) = x_{AB}^*(r^* = 1) = x_{AB}(r^* = 1)/x_{AB}(\infty)$ at the surface of the reaction zone is plotted in Fig. (3.3.5). This molar ratio affects the reaction rate as seen from Eq. (3.3.15). It is seen from these plots that for negative α_T^* values the molar ratio can increase, thus causing an increase in the reaction rate. This was caused by the increase in the gas temperature in regions outside the reaction zone as seen in Fig. (3.3.4) which increased the temperature-dependent diffusion coefficient, D_{AB} [$D_{AB}(T) = D_{AB}(\infty)T^n(r^*)/T(\infty)$], towards the surface of the reaction zone. Additional results indicated that the effects of thermal diffusion and chemical convection on the temperature distribution in the ambient becomes less important with decreasing surface temperature, T_s^* [$T_s^* = T(r_D)/T(\infty)$] and almost vanishes for $T_s^* \leq 5$.

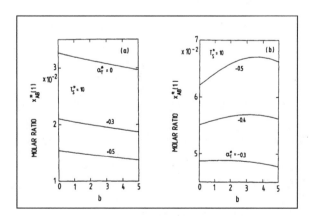

Fig. 3.3.5. Molar ratio at the deposit surface, $x_{AB}^*(1)$, as a function of b for surface temperature, $T_s^* = 10$ and different α_T^* values. (a) $\alpha_T^* \geq 0$, (b) $\alpha_T^* < 0$ (reprinted from Luk'yanchuk et al. 1992 with permission from Elsevier Science)

In laser chemical vapor deposition, reactions occur mainly on the solid surface (heterogeneous reactions). However, reactions can occur within the gas phase above the surface (homogeneous reactions). A model considering the effects of heterogeneous and homogeneous reactions on the reaction rate has been considered by Kirichenko and Baeuerle (1992). The model assumes a semi-infinite substrate irradiated by a continuous laser beam perpendicular to the surface. The laser light is absorbed only by the substrate and not the reactive gas. Gas-phase heating is caused only by heat flux from the laser heated solid surface. The reaction type considered has the general form

$$\zeta_{AB} AB + \zeta_c C + G \overset{k_1, k_2, k_3}{\rightarrow} \zeta_A A_s + \zeta_{BC} BC + G, \qquad (3.3.31)$$

where the $\zeta_i{'}$s are the stoichiometric coefficients, k_1 and k_3 are the reaction rate constants for the surface and gas phase forward reactions, respectively; k_2 is the rate constant for the backward reaction; and G is the inert gas carrier. The reaction rate constants are temperature-dependent and are described by the Arrhenius equation. If A is generated in the gas-phase, it must diffuse to the surface to form the solid deposit, A_s. The interest here is in the forward reaction. The backward reaction is pure etching. Assuming cylindrical symmetry, the steady state heat equation describing the temperature distribution $T(r, z)$ in the solid ($z \leq 0$) or gas phase ($z > 0$) is given by

$$\frac{1}{r}\frac{\partial}{\partial r}(r\frac{\partial T}{\partial r}) + \frac{\partial^2 T}{\partial z^2} = 0 \ ; \ z \leq 0 \text{ or } z > 0. \qquad (\,3.3.32)$$

Here, the boundary conditions are

$$T(r, z \rightarrow \pm\infty) = T(\infty); \ T(z = -0) = T(z = +0), \qquad (3.3.33)$$

and

$$-\kappa_s \frac{\partial T}{\partial z}\mid_{z=-0} = -\kappa_g \frac{\partial T}{\partial z}\mid_{z=+0} -I(r), \qquad (3.3.34)$$

where $I(r) = I(0)\exp(-\frac{r^2}{\omega_o^2})$ is the Gaussian beam intensity with radius ω_0..

For determining the spacial concentration of species AB, BC, and A, thermal diffusion and chemical convection were ignored. As such, for species AB, one has

$$\nabla(ND_{AB}\nabla x_{AB}) = W_g; \ x_{AB}(r, z \rightarrow \infty) = x_{AB}(\infty); \text{ and}$$

$$-J_{AB} = ND_{AB}\frac{\partial x_{AB}}{\partial z}\mid_{z=0} = W_s, \qquad (3.3.35)$$

where W_s and W_g are the reaction rates at the surface and in the gas-phase, respectively. For species BC,

$$\nabla(ND_{BC}\nabla x_{BC}) = -W_g \ ; \ x_{BC}(r, z \rightarrow \infty) = 0, \text{ and}$$

$$ND_{BC}\frac{\partial x_{BC}}{\partial z}\mid_{z=0} = -W_s, \qquad (3.3.36)$$

and for species A,

$$\nabla(ND_A\nabla x_A) = -W_g; \quad x_A(r, z \to \infty) = 0; \quad x_A(z = 0) = 0. \tag{3.3.37}$$

Approximate solutions were obtained for Eqs. (3.3.32) and (3.3.35)-(3.3.37) and applied to special reaction cases (for details see Kirichenko and Baeuerle 1992). Of interest are pure gas-phase reaction ($k_3 \neq 0, k_1 = k_2 = 0$), pure surface reaction ($k_1 \neq 0, k_2 = k_3 = 0$), and a combination of both. Fig. (3.3.6) shows the normalized reaction rate as a function of the normalized distance, $r^* = r/w_0$ for the case of gas phase reaction with $E_a/k_B = 8,000K$. The solid curves represent different temperatures on the surface at the center of the beam ($r = 0, z = 0$). The dotted curve represents the normalized distribution of the laser intensity. It is seen that as the temperature at the center of the beam increases the radial distance of the deposit increases. This is explained by the fact that an increase in temperature increases the height of the gas phase reaction layer over the irradiated surface which allows the A atoms to diffuse over a larger distance to the surface. As a result, A is deposited over a wider area leading to an increase in the radial distance of the deposit. The shape of the deposit is quite similar to the Gaussian shape of the beam intensity. A similar plot representing the normalized reaction rate as a function of the normalized radial distance for pure surface reaction is shown in Fig. (3.3.7). Here, $E_a/k_B = 20,000K$. Solid curves represent different center temperatures, and the dotted curve is the beam normalized distribution of the laser intensity. Here, the deposit seems more localized in the radial direction than in the case of pure gas phase reaction. For a high temperature of $2,300K$, the deposit exhibits a crater or volcano-like shape at the center. This could be caused by the transport limitation of species AB in the center region of the reaction zone. No new results were observed when gas phase and surface reactions were considered simultaneously.

The kinetic effect on the deposition rate through the Arrhenius equation may be dependent on the precursor gas concentration as determined by gas diffusion. In this case, the growth rate becomes affected by the gas mass transfer and the surface temperature. Such a model has been proposed by Conde et al. (1992) in conjunction with the deposition of TiN dots from a gas mixture of $TiCl_4$, H_2, and N_2. The reaction scheme by which TiN is deposited on the surface has been described by the expression $TiCl_4(g) + 2H_2(g) + \frac{1}{2}N_2(g) \to TiN(s) + 4HCl(g)$, and the rate of deposition of TiN is given by

$$R_{TiN} = \left[C_{H_2}\sqrt{C_{N_2}}C_{TiCl_4}\right]k_0' \exp\left(-\frac{E_a}{R_gT_s}\right), \tag{3.3.38}$$

where R_{TiN} is the mass of TiN deposited per unit area per unit time; C_{H_2}, C_{N_2}, and C_{TiCl_4} are concentrations, measured as mass per unit volume, for each of the gas species at any point (x, y, z) at any time t, and $T_s = T(x, y, 0, t)$ is the surface temperature at any point (x, y). In applying Eq. (3.3.38), C_{H_2}, and C_{N_2} were set equal to their initial values $(CH_{2_0}, CN_{2_0}^{1/2})$

which meant that they were considered as constants This was justified by the
fact that they were more than an order of magnitude higher than the $TiCl_4$
mole fraction in the experimental study. As such, Eq. (3.3.38) can be expressed
as

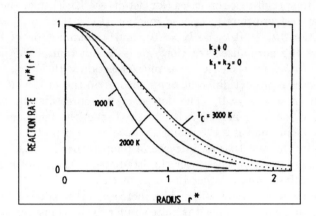

Fig. 3.3.6. Reaction rate (normalized) as a function of the normalized radial
distance, r^*, for the case of pure gas-phase reaction and $\frac{E_a}{k_B} = 8000K$. The solid
curves represent different surface center temperatures, T_c, and the dotted curve
the normalized laser-intensity distribution (reprinted from Kirichenko and
Baeuerle 1992 with permission from Elsevier Science)

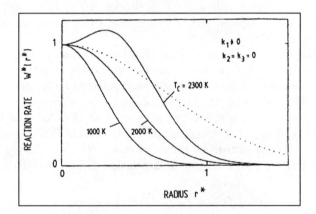

Fig. 3.3.7. Normalized reaction rate for pure surface reaction as a function of r^*.
$\frac{E_a}{k_B} = 20000K$, and $T_0 = 2000K$. The solid curves represent different surface
temperatures at the center, T_c, and the dotted curve the normalized laser-intensity
distribution (reprinted from Kirichenko and Baeuerle 1992 with permission from
Elsevier Science)

$$R_{TiN} = AC_{TiCl_4} k_0 \exp\left(-\frac{E_a}{R_g T_s}\right), \tag{3.3.39}$$

where $A = CH_{2_0} \times CN_{2_0}^{1/2}$. It is clear from Eq. (3.3.39) that the rate of deposition of TiN is a function of the surface temperature and the concentration of $TiCl_4$. Hence, the thickness of the deposit can be determined by determining the surface temperature, T_s, from the three-dimensional heat conduction equation and the $TiCl_4$ concentration, C, from the three-dimensional mass diffusion equation

$$\frac{\partial C}{\partial t} = D\left(\frac{\partial^2 C}{\partial x^2} + \frac{\partial^2 C}{\partial y^2} + \frac{\partial^2 C}{\partial z^2}\right), \tag{3.3.40}$$

where D is the diffusion coefficient of $TiCl_4$.

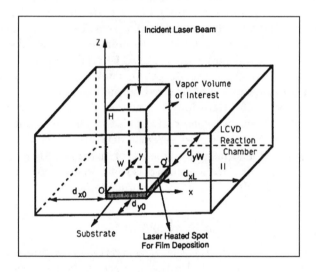

Fig. 3.3.8. Schematic diagram of the LCVD reactor and inner rectangular region used in the model (reprinted from Conde et al. 1992 with permission from the American Institue of Physics)

Both the heat equation and Eq. (3.3.40) were solved over the rectangular region shown in Fig. (3.3.8). Eq. (3.3.40) was solved using the double integral Fourier transform in the x and y directions. Model simulation of deposit thickness as a function of the distance x from the beam center was compared to experimental observations for various parameters. For laser powers of 400 W and 600 W, the surface was crater or volcano in shape in the central region of the beam in agreement with experimental observations. This volcano surface was produced by assuming a sticking or an adsorption coefficient for

$TiCl_4$, which (within a certain temperature range, $T_m - T_M$) decreased linearly with an increase in temperature. As such, the adsorption coefficient was at a minimum in the central region, where the temperature was at a maximum. Figures (3.3.9) and (3.3.10) demonstrate the volcano shape deposit at the center. Parameters employed for Fig. (3.3.9) were beam radius = $1.1mm$, $T_m = 1473K$, $T_M = 1640K$, $E_a = 12.2\ kcal/mol$, and $k_0' = 6.9 \times 10^2 cm/s$. Those employed for Fig. (3.3.10) were the same as for Fig. (3.3.9) except for k_0' = $Ak_0 = 7.9 \times 10^2\ cm/s$. It is clear from these figures that the model showed good qualitative agreement with experimental observations, but varied in how well it agreed quantitatively with observed data.

All models presented so far are based on a continuous laser source (CW laser-irradiation). A different problem arises when a pulsed laser is applied as a heat source. In short pulsed laser applications, effects caused by the dark period between pulses become important. Konstantinov et al. (1990) proposed a one-dimensional diffusion model to study the effect of gas phase transport on the deposition or growth rate in a pulsed-laser chemical vapor deposition system. The simplifying assumptions involved were as follows. (1) The intensity distribution of the beam is assumed to be uniform on a circular-cylindrical domain whose axis is defined by the z-axis. The origin of the coordinate system is at the surface of the substrate. All precursor gas molecules within the confines of the laser beam are photolyzed during one pulse. (2) The duration of the light period (pulse duration) is much smaller than that of the dark period (interval between two consecutive pulses). This pulse interval is $1/f$, where f is the pulse repetition rate. As such, deposition occurs mainly during the pulse interval. (3) The gas phase consists of molecules AB in a carrier gas. Upon laser radiation the AB molecules dissociate to give molecules A and B. (4) Molecules A diffuse to the surface to form the deposit. There is a loss of the A molecules described by the expression

$$n' = n\exp(-t/\tau), \tag{3.3.41}$$

where τ is the mean lifetime, n' is the concentration of species A contributing to the growth of the deposit, and n is the concentration of species A in the gas phase. The pressure of the buffer or carrier gas is much higher than the partial pressure of the A molecules in which case one may approximate the mass transport by a one-dimensional diffusion process along the z axis. (5) The characteristic diffusion length is much smaller than the size of the deposition cell, which justifies using asymptotic boundary conditions. (6) Deposition is caused by the diffusion flux of species A at the surface. The concentration n' can be written in the form

$$\frac{\partial n'}{\partial t} = D\frac{\partial^2 n'}{\partial z^2} + \frac{n'}{\tau}. \tag{3.3.42}$$

Substituting Eq. (3.3.41) into Eq. (3.3.42) gives the equation

$$D\frac{\partial^2 n}{\partial z^2} = \frac{\partial n}{\partial t}. \tag{3.3.43}$$

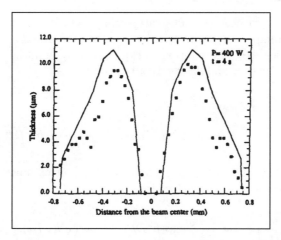

Fig. 3.3.9. Thickness of TiN deposits as a function of distance from the beam center (reprinted form Conde et al. 1992 with permission from the American Institute of Physics)

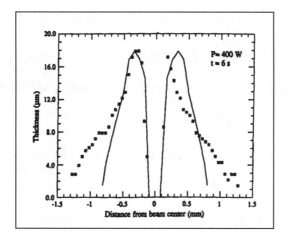

Fig. 3.3.10. Thickness of TiN deposit as a function of distance from the beam center (reprinted from Conde et al. 1992 with permission from the American Institute of Physics)

Here, the initial and boundary conditions are expressed as

$$n(z,0) = n_0 \; ; \; n(\infty,t) = n_0 \; ; \; \text{and} \quad D\left(\frac{\partial n}{\partial z}\right)_{z=0} = \eta v n(0,t)/4. \quad (3.3.44)$$

In Eqs. (3.3.42) and (3.3.44), D is the diffusion coefficient for species A, η is the sticking coefficient at the surface, v is the mean velocity of the A molecules

$(v = (8kT/\pi m)^{1/2})$ near the substrate surface, where k is the Boltzmann's constant, T is temperature in Kalvin, m is particle mass, and n_0 is the concentration of species AB. The solution to Eq. (3.3.43) with initial and boundary conditions is (Carlslaw and Jaeger 1959)

$$n'(z,t) = n_0 \exp(-t/\tau)[1 - \operatorname{erf} c(\frac{z}{2\sqrt{Dt}}) \tag{3.3.45}$$
$$+ \exp(hz + h^2 Dt)\operatorname{erf} c(\frac{z}{2\sqrt{Dt}}) + h\sqrt{Dt})],$$

where $h = \eta v/4D$ and $erfc$ is the complementary error function $(1 - erf)$. The flux of species A at the surface is

$$D\left(\frac{\partial n'}{\partial z}\right)_{z=0} = \frac{\eta v n_0}{4} \exp(h^2 Dt - t/\tau)\operatorname{erf} c(h\sqrt{Dt}). \tag{3.3.46}$$

Growth rate of the deposit per pulse, R (number of atoms of A deposited on a unit area per unit time) is obtained by integrating over the pulse interval $(0 - 1/f)$. This gives

$$R = \frac{n_0 \eta \tau v}{4(y-1)}[\exp(xy - x)\operatorname{erf} c(\sqrt{xy}) + \sqrt{y}\operatorname{erf}(\sqrt{x}) - 1], \tag{3.3.47}$$

where $x = (\tau f)^{-1}$, and $y = \eta^2 v^2 \tau/16D$. The concentration n_0 was assumed constant over the short time pulse interval. However, over a long time period, it was assumed that n_0 changes with time. The equation describing this change is expressed as

$$V\frac{dn_0}{dt} = qn_{00} - qn_0 - V_{ph}fn_0, \tag{3.3.48}$$

where $n_{00} = n_0(0)$, q is the flow rate into the deposition cell, V_{ph} is the volume of the photolyzed molecules, and V is the volume of the deposition cell $(V \gg V_{ph})$. Solution of Eq. (3.3.48) gives

$$n_0(t) = \frac{n_{00}}{1+\mu}[1 + \mu\exp(-q(1+\mu)t/V)], \tag{3.3.49}$$

where $\mu = V_{ph}f/q$. Hence the growth rate through Eqs. (3 3.47) and (3.3.48) is now dependent on the pulse repetition rate, f, and on time, t. For realistic parameter values, the second term in brackets in Eq. (3.3.52) dominates. Hence, from Eq. (3.3.47) and the long-term limiting case of Eq. (3.3.49), one obtains an expression relating growth rate to the pulse repetition rate.

$$R \propto \operatorname{erf}(1/\sqrt{\tau f})[1 + V_{ph}f/q]^{-1}. \tag{3.3.50}$$

Using the same argument in deriving Eq. (3.3.50) and assuming that τ and μ are proportional to pressure, one obtains an expression relating the growth rate to the buffer gas pressure.

$$R \propto (1 + \mu_0 p/p_0)^{-1} \, \mathrm{erf}(1/\sqrt{f\tau_0 p/p_0}). \qquad (3.3.51)$$

The dependence of R on the gas flow rate is obtained also from Eq. (3.3.50) for a given f.

$$R \propto [1 + V_{ph} f/q]^{-1}. \qquad (3.3.52)$$

Equations (3.3.50)-(3.3.52) were fitted to experimental data on growth rate as a function of (1) pulse repetition rate, (2) total pressure, and (3) gas flow rate. Experimental data were obtained from the deposition of Cr from $Cr(CO)_6$ in an argon buffer (Ar) using a K_rF excimer laser. The fit of the model to the data was rather good. However, some discrepancies were observed between model parameters as estimated from the fit of the model to the data and the same parameters determined experimentally. The model certainly gave good qualitative agreement with experimental data. Results showed that for a fluence of $80mJ/cm^2$, an argon flow rate of 35 sccm, and a total pressure of 80 mbar, the growth rate (A^0 per pulse) decreased non-linearly with an increase in the pulse repetition rate (pps) in the range of 5 to 50 pulses per second. The maximum growth rate was about 3 A^0 per pulse at about 5 pps. For 5 pps pulse repetition rate, $110mJ/cm^2$, and 50 sccm Ar flow rate, growth per pulse decreased non-linearly with an increase in pressure, from about 30 mbar to about 700 mbar. The maximum growth rate of about 55 A^0 occurred at the lowest pressure. Also, for $62\ mJ/cm^2$ and 5 pps, growth rate (multiplied by the optical absorption coefficient of the deposit film) increased almost linearly with an increase in flow rate (in the range 15 to 50 sccm) for 200 and 370 mbar pressures, but growth was nonlinear at a low pressure of 75 mbar. As expected, maximum growth rate occurred at the 75 mbar and 50 sccm factor combination.

4

Laser Photopolymerization

4.1 Introduction

laser-induced photopolymerization is a new rapid prototyping technique that can be used in free-form manufacturing of high aspect ratio microstructures. The process consists of UV light-induced ultrafast polymerization of multi-functional monomers. The prototype to be manufactured is built layer by layer in a stepwise manner. The system is composed of a laser, a multiaxial positioning stage with a stage-motion controller, and a computer control having interfaces with the optical and mechanical part of the system. A database describing a three-dimensional microstructure of specified dimensions and topology is generated by a computer software code (AutoCAD). The structure is then sliced into thin two-dimensional layers. This information is utilized in the computer to position the laser over a thin layer of liquid photo-sensitive monomers on top of a platform immersed in a vat of liquid monomers. The thin layer of monomers is polymerized and solidified when exposed to the laser light. Laser exposure can be free-form or through a mask. In free-form exposure, scanning of the monomer layer by a pulsed laser beam is accomplished through computer-controlled movement of the stage in the x,y plane that coincides with the shape of the thin layer generated by the software code. The solidified layer residing on the platform is lowered to allow another thin layer of liquid monomers on top of the solidified layer (another technique raises a UV window positioned on top of the solidified monomer layer to allow a new layer of monomer liquid to flow on top of the solid layer). The irradiation process is then repeated and layers are stacked until the prototype microstructure is produced.

This photoforming system is capable of producing three-dimensional microstructures of various shapes and dimensions, is automatically controlled and simple to execute, and is fast and inexpensive compared to other techniques such as x-ray lithography. Dimensional accuracy and surface smoothness of the photopolymerized microstructure depend on the kinetics and uniformity of monomer conversion in forming the solidified layers, on the monomer layer thickness, on the laser light intensity and beam width, and on the scanning speed and scanning pattern (i.e., the effective laser dwell time or the number of pulses at each pixel on the surface) of the laser beam on the monomer layer surface. To achieve a system with high resolution, accurate and predictive mathematical models of the focused laser beam photopolymerization process must be developed and utilized in the system control scheme. In this chapter, we discuss kinetics of photo-initiated free radical polymerization and models to predict monomer conversion and solid formation.

4.2 Kinetics of Photopolymerization

Successful modeling of a photopolymerization microfabrication system depends on knowledge of the kinetics of photo-initiated free radical polymerization. A kinetic study of acrylate photopolymerization to determine rate constants was undertaken by Tryson and Shultz (1979). Free radical polymerization can be described in terms of the following steps (Oster and Yang, 1968):

$$I \rightarrow 2R_*,$$

$$R_* + M \xrightarrow{k_p} RM_*,$$

$$RM_{n*} + M \xrightarrow{k_p} RM_{n+1*}, \tag{4.2.1}$$

$$RM_{n*} + RM_{m*} \xrightarrow{k_t} \text{terminated polymer},$$

where I is the initiator species which yields free radicals (R_*) upon laser exposure, M is a monomer, k_p and k_t are reaction rates for propagation and termination, and RM_{n*} is a chain radical. Free radicals are formed upon illumination at the rate ΦI_a, where Φ is the quantum yield (the number of propagating chains produced per photon absorbed) and I_a the rate of photon absorption. Here, chain termination is assumed to be bimolecular (reaction of two radical chains) and is due to combination or disproportionation. If the sum of concentrations of all RM_{n*} chains, including R_*, is denoted by $[P_*]$, then under steady state conditions one has

$$\frac{d[P_*]}{dt} = 0 = \Phi I_a - k_t[P_*]^2 \tag{4.2.2}$$

or

$$[P_*] = (\Phi I_a/k_t)^{1/2}. \tag{4.2.3}$$

the rate of change of the monomer concentration through conversion is seen to be

$$-\frac{d[M]}{dt} = k_p[P_*][M]. \tag{4.2.4}$$

replacing $[P_*]$ by its value from Eq. (4.2.3), gives

$$-\frac{d[M]}{dt} = k_p(\Phi I_a/k_t)^{1/2}[M].$$ (4.2.5)

From Eq.(4.2.5), it is seen that

$$\frac{k_p}{(k_t)^{1/2}} = \frac{-d[M]/dt}{(\Phi I_a)^{1/2}[M]}.$$ (4.2.6)

The right-hand side of Eq. (4.2.6) is determined from differential scanning calorimetry (DSC) data, since $-d[M]/dt$ is related to the exotherm rate, the monomer concentration $[M]$ is determined from monomer conversion, and ΦI_a is obtained from light intensity measurements. Here, the absorbed light intensity at depth z in average number of photons per cm^3 per sec is $I_a = I_0 z^{-1}(1 - \exp(-2.303\varepsilon cz))$, where c is the photo-initiator concentration and ε is the molar extinction coefficient.

When there is no light ($I_a = 0$), Eq.(4.2.2) yields

$$\frac{d[P_*]}{dt} = -k_t[P_*]^2.$$ (4.2.7)

At $t = 0$, one has from Eq. (4.2.4) the initial condition given by

$$[P_*]_0 = \frac{(-d[M]/dt)_0}{k_p[M]_0}.$$ (4.2.8)

Solving Eq.(4.2.7) for $[P_*]$ yields

$$\frac{1}{[P_*]} = k_t t + \frac{1}{[P_*]_0}.$$ (4.2.9)

Replacing $[P_*]$ by $-\frac{(d[M]/dt)}{k_p[M]}$ from Eq.(4.2.4) and substituting $[P_*]_0$ from Eq.(4.2.8), one obtains

$$\frac{[M]}{(-d[M]/dt)} = \frac{k_t}{k_p}t + \frac{[M]_0}{(-d[M]/dt)_0}.$$ (4.2.10)

Values for $[M]$ and $d[M]/dt$ can be obtained from DSC experimental data over time. Hence, $\frac{k_t}{k_p}$ can be calculated from the slope to the least square line expressing $\frac{[M]}{(-d[M]/dt)}$ as a function of time. Using estimates of $\frac{k_t}{k_p}$ from Eq.(4.2.10) and of $\frac{k_p}{(k_t)^{1/2}}$ from Eq. (4.2.6), one may obtain individual estimates of k_t and k_p. These estimates are tabulated (Table 4.2.1) for the monomers 1,6-hexanedioldiacrylate ($HDDA$) and pentaerythritol tetraacrylate (PET_4A) at different percent monomer conversion. Benzoin ethyl ether (BEE) was used as the photo-initiator.

In general, results in Table 4.2.1 indicate a decrease in the rate constants with an increase in percent conversion for both multifunctional acrylates. Also, it is observed that k_t decreases at a faster rate than k_p with an increase in conversion. Polymerization using these multifunctional acrylates involves network formation which increases with an increase in monomer conversion. As a result, radicals as well as monomers become more entangled in the network as it increases in formation. This entanglement inhibits the diffusion of monomers to reactive sites which results in decreasing k_p. Similarly, radicals are inhibited from diffusing together which causes a decrease in k_t. Hence, the decrease in both k_t and k_p can be attributed to diffusion limitation caused by an increase in network formation at higher percent monomer conversion.

Table 4.2.1. Rate constants for $HDDA$ at $30°C$ and 0.239% by weight BEE and for PET_4A at $40°C$ and 0.981% by weight BEE (reproduced from Tryson and Shultz 1979 by permission of John Wiley and Sons, Inc.)

	HDDA			PET$_4$A	
Conversion	$k_p \times 10^{-2}$	$k_t \times 10^{-4}$	Conversion	k_p	$k_t \times 10^{-2}$
$\%$	$l\,mol^{-1}\,sec^{-1}$	$l\,mol^{-1}\,sec^{-1}$	$\%$	$l\,mol^{-1}\,sec^{-1}$	$l\,mol^{-1}\,sec^{-1}$
6.9	140	640	1	130	970
11.8	290	1300	1.2	150	850
18.8	270	590	3.8	340	510
24.6	190	200	6.2	170	240
33.6	63	23	9.9	64	77
38.3	46	15	11	47	61
44.1	21	5	14.4	32	7
49.1	9	1.8	21.3	6.3	2.9

A free-radical polymerization model to account for diffusion was formulated by Achilias and Kiparissides (1992). In this model, initiator efficiency as well as propagation and termination rate constants were expressed in terms of a reaction-limited term and a diffusion-limited term. The initiator efficiency factor, f $(0 < f \leq 1)$, which represents the fraction of generated primary radicals that leads to the formulation of new polymer chains, can decrease with monomer conversion indicating a diffusion-limitation problem. Two concentric spheres of radius r_1 and r_2 $(r_1 < r_2)$ were assumed to comprise the sample space of primary radicals. These radicals are generated by the initiator reaction within the sphere of radius r_1. They must diffuse through the sphere of radius r_2 before they can react with monomers to initiate new polymer chains. It was shown that

$$\frac{1}{f} = \frac{1}{f_0} + \frac{r_2^3 k_i [M]}{3 r_1 f_0 d_R}, \tag{4.2.11}$$

where f_0 is the initiator efficiency at time $t = 0$, k_i is the rate constant for the reaction of a primary radical with a monomer to initiate polymer chain, and d_R is the diffusion coefficient of the primary radicals. An expression for the diffusion-limited termination rate constant is given by

$$\frac{1}{k_t} = \frac{1}{k_{t_0}} + \frac{r_t^2 \lambda_0}{3d_{pe}}, \qquad (4.2.12)$$

where r_t is a radius defining the reaction spherical volume within which polymer termination occurs, k_{t_0} is an intrinsic termination rate constant, λ_0 is the total concentration of radical chains, and d_{pe} is the effective diffusion coefficient of the polymer and is equal to the probability that two polymers in close proximity react multiplied by the diffusion coefficient of the polymer, d_p. A functional relationship for the diffusion-controlled propagation rate constant is expressed as

$$\frac{1}{k_p} = \frac{1}{k_{p_0}} + \frac{r_m^2 \lambda_0}{3d_m}, \qquad (4.2.13)$$

where k_{t_0} is an intrinsic propagation rate constant, r_m is a reaction radius, and d_m is the monomer diffusion coefficient.

The kinetic steps in free-radical polymerization are basically an extension of Eq. (4.2.1) which include initiation, propagation, chain transfer to monomer, and termination. These steps can be expressed as
Initiation

$$I \xrightarrow{k_d} 2R_*$$

$$R_* + M \xrightarrow{k_i} RM_*,$$

propagation

$$RM_{n*} + M \xrightarrow{k_p} RM_{n+1*},$$

chain transfer to monomer

$$RM_{n*} + M \xrightarrow{k_f} RM_* + M, \qquad (4.2.14)$$

termination by combination

$$RM_{n*} + RM_{m*} \xrightarrow{k_{t_c}} D_{n+m},$$

and termination by disproportionation

$$RM_{n*} + RM_{m*} \xrightarrow{k_{t_d}} D_n + D_m,$$

where the $k's$ are the rate constants.

Based on these kinetic steps, one can derive differential equations to simulate the evolution of the polymerization reaction and molecular weight. For a well-mixed batch reactor of volume V, these equations may be expressed as

$$\frac{d[I]}{dt} = -k_d[I] - \frac{[I]}{V}\frac{dV}{dt},$$

$$\frac{d[R_*]}{dt} = 2fk_d[I] - k_i[R_*][M] - \frac{[R_*]}{V}\frac{dV}{dt},$$

$$\frac{d[M]}{dt} = -k_i[R_*][M] - (k_p + k_f)[M]\lambda_0 - \frac{[M]}{V}\frac{dV}{dt},$$

$$\frac{dV}{dt} = -V_0\epsilon\{k_i[R_*] + (k_p + k_f)\lambda_0\}\frac{[M]V}{[M_0]V_0},$$

$$\frac{d\lambda_0}{dt} = k_i[R_*][M] - k_t\lambda_0^2 - \frac{\lambda_0}{V}\frac{dV}{dt}. \tag{4.2.15}$$

Here, ϵ is the volume contraction factor defined as $(\rho_p - \rho_m)/\rho_p$, where the $\rho's$ represent the densities of polymer and monomer. Also, λ_0 is the total concentration of radical chains. Monomer conversion is defined as $X = (M_0V_0 - MV)/M_0V_0$.

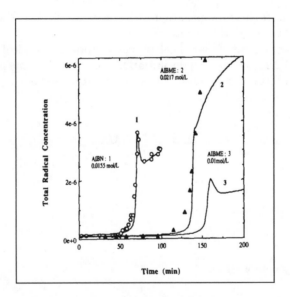

Fig. 4.2.1. Total radical chain concentration (λ_0) as a function of time for the polymerization of MMA with different initiator types and concentrations. Data Points: (o) from Zhu et al. 1990 with; permission from Elsevier Science; (▲) from Shen et al. 1987 with permission from Wiley (reprinted from Achilias and Kiparissides 1992 with permission from the American Chemical Society)

The model equations were solved numerically and used to predict monomer conversion and total radical chain concentration as a function of time. Predicted values were in good agreement with the experimental values for methyl methacrylate (MMA) and styrene polymerization. Numerical values for the parameters in the model were obtained independently of the experiments. Figure (4.2.1) presents total radical chain concentration as a function of time for MMA and different initiators. The solid curves are predicted from the model and the points are observed data. There is good agreement between predicted and observed values. Figure (4.2.2) also shows good agreement between observed and predicted values for monomer conversion over time.

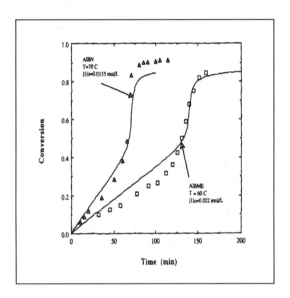

Fig. 4.2.2. Monomer conversion as a function of time for MMA polymerization with two types of initiators. Data points: (▲) from Zhu et al. 1990 with permission from Elsevier Science; (□) from O'Driscoll and Huang 1989 with permission from Elsevier Science (reprinted from Achilias and Kiparissides 1992 with permission from the American Chemical Society)

A photopolymerization model which considered, in addition to diffusion-limited propagation and termination, a so-called reaction diffusion termination mechanism was developed by Anseth and Bowman (1993). Diffusion-controlled termination is comprised of three steps: translational diffusion (diffusion of the growing polymer chain), segmental diffusion (involves a re-orientation of the polymer segments so that the polymer chain ends are within a reaction zone), and reaction between two radical ends leading to termination. At higher monomer conversion, the reacting medium increases in viscosity, and translational diffusion decreases to the point where it becomes insignificant. Under these conditions, an immobile radical chain grows until it encounters a second radical for termination (in this case, the only possible movement of the radical chain is through propagation reaction). This termination mechanism has been termed reaction diffusion termination. Propagation remains fairly constant at low conversion. At high conversion, it deceases because of the lower mobility of monomers. In developing their model, Anseth and Bowman (1993) incorporated the concepts of free volume and volume relaxation, ignored translational diffusion as being insignificant for highly cross-linked polymers, neglected changes in the initiator efficiency, and considered only termination by the combination of two radicals. The model was a modification of those previously developed by Marten and Hamielec (1979) and Bowman and Peppas (1991) to include reaction diffusion-controlled termination. Expressions were derived for the propagation and termination constants as functions of free volume and critical volumes for propagation and termination. The propagation constant is given by

$$k_p = k_{p0} \frac{1}{1 + \exp[B(1/v_f - 1/v_{f,cp})]},$$ (4.2.16)

and the termination constant by

$$k_t = k_{t_0}[1 + [\frac{1}{R\{\frac{1}{1+\exp[B(1/v_f-1/v_{f,cp})]}\} + \exp\{-A(1/v_f - 1/v_{f,ct})\}}]]^{-1},$$ (4.2.17)

where A, B, and R are constants, k_{p0} and k_{t_0} are the initial rate constants, v_f is the free volume, $v_{f,cp}$ is the critical free volume where transition occurs from reaction-controlled to diffusion-controlled propagation, and $\nu_{f,ct}$ is the critical free volume for segmental-controlled termination. The free volume is expressed as

$$v_f = v_{f,eq} + \frac{v - v_\infty}{v_\infty},$$ (4.2.18)

where v is the specific volume and v_∞ is the equilibrium specific volume. Here, $v_\infty = \nu_m(1 - X\xi_v)$, where ν_m is the specific volume of the monomer and ξ_v is the volume contraction factor. The equilibrium free volume is

$$v_{f,eq} = 0.025 + \alpha_m(T - T_{g,m})(1 - \phi_p) + \alpha_p(T - T_{g,p})\phi_p,$$ (4.2.19)

where 0.025 is the assumed fractional free volume at the glass transition temperature, α is the thermal expansion coefficient, T is the reaction temperature, $T_{g,m}$ and $T_{g,p}$ are the glass transition temperatures for monomer and polymer, respectively, and ϕ_p is the polymer volume fraction. The latter is given by

$$\phi_p = \frac{X(1 - \varepsilon_v)}{1 - X\varepsilon_v},$$ (4.2.20)

where X is monomer conversion. Figures (4.2.3) and (4.2.4) present the propagation (k_p) and termination (k_t) constants as a function of conversion for the photopolymerization of diethylene glycol dimethacrylate (DEGDMA). Model parameters used for the simulation are given in Table 1 of Anseth and Bowman (1993). It is seen from Fig.(4.2.3) that k_p remained relatively constant up to a conversion in the neighborhood of 0.2 after which it began to decrease rapidly. The termination kinetic constant k_t in Fig.(4.2.4) exhibits a different behavior. Termination decreases from the onset of polymerization (region I of the curve representing segmental diffusion control). In region II, which represents reaction diffusion control, termination remains constant as long as the propagation rate is constant. In this region, termination is seen to plateau somewhat. Region III represents monomer diffusion-controlled propagation. It is seen in this region that termination decreases as a result of propagation decrease caused by monomer diffusion control. These results are in qualitative agreement with observed experimental results by Garrett et al. (1989) and with predicted results in Fig.(4.2.5) from Achilias and Kiparissides (1992) based on different model equations and on the polymerization of MMA with AIBN initiator. The quantitative changes in k_p and k_t in Fig.(4.2.5) are quite different from those seen in Figs. (4.2.3) and (4.2.4). In Fig. (4.2.5), the propagation rate remains constant over a wide range of monomer conversion, and the decrease in termination rate, through four different stages, is not as drastic as that in Fig. (4.2.4). This may be due to a difference in models and/or to different monomers used for polymerization. However, as stated by Anseth and Bowman (1993), their model (with a limited number of fitting parameters) was intended for predicting polymerization behavior more qualitatively than quantitatively. It has the advantage of being relatively simple and easy to apply.

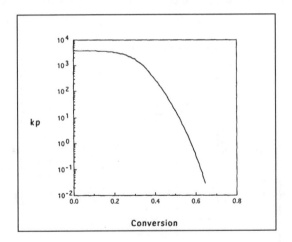

Fig. 4.2.3. Propagation rate constant as a function of monomer conversion (reprinted from Anseth and Bowman 1993, p 512, by courtesy of Marcel Dekker Inc.)

In a similar kinetic modeling approach to Achilias and Kiparissides (1992), and Anseth and Bowman (1993), Kurdikar and Peppas (1994) developed a diffusion-controlled cross-linked photopolymerization model by considering changing initiator efficiency with conversion, diffusion-limited propagation and termination as well as volume relaxation. The propagation rate was considered to be a function of translational and segmental diffusion and given by

$$\frac{1}{k_p} = \frac{1}{k_{p,trans}} + \frac{1}{k_{p,seg}}, \tag{4.2.21}$$

where (Rice, 1985)

$$k_{p,trans} = 4\pi(r_m + r_r)D_t, \tag{4.2.22}$$

and

$$k_{p,seg} = 4\pi(r_m + r_r)D_s. \tag{4.2.23}$$

Here, r_m and r_r are the reaction radii of the monomer and radical, and D_t and D_s are the translational and segmental diffusion coefficients of the monomer. The translational diffusion coefficient D_t was calculated from Vrentas and Duda (1977a,b) for linear-solvent polymer systems (Eq. (13) of Kurdikar and Peppas 1994). The relaxation volume (defined as the deviation of the specific volume from the equilibrium specific volume) was included in the calculation of the translational diffusion of the monomer. The segmental diffusion coefficient D_s was set equal to pD_t, where p is the probability that a double bond of the monomer is oriented toward the radical site (Eq. (27) of Kurdikar and Peppas 1994). The initiator efficiency is given in Eq. (8) of Kurdikar and Peppas. Termination was considered for the case of reaction diffusion control. An expression for k_t was given by

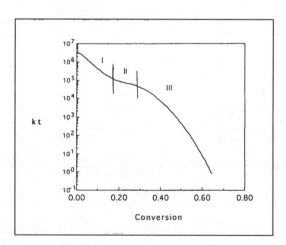

Fig. 4.2.4. termination rate constant as a function of conversion (reprinted from Anseth and Bowman 1993, p 513, by courtesy of Marcel Dekker Inc.)

$$k_t = \frac{32}{3}\pi r_r l^2 k_p[M], \qquad (4.2.24)$$

where l is the bond distance.

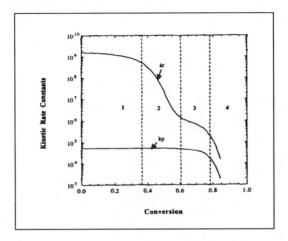

Fig. 4.2.5. Propagation and termination rates as a function of conversion (reprinted from Achilias and Kiparissides 1992 with permission from the American Chemical Society)

Assuming the kinetic steps in Eq. (4.2.1), the rate of change of radical concentration can be expressed as

$$\frac{d[P_*]}{dt} = 2f\phi' I_0\epsilon[A] - 2k_t[P_*]^2, \qquad (4.2.25)$$

where f is the initiator efficiency (defined as the fraction of radicals produced that initiate propagation chains), ϕ' is the number of initiator molecules produced per absorbed photon ($\phi = f\phi'$), ϵ is the extinction coefficient of the initiator, $[A]$ is the concentration of the initiator, and I_0 is the incident light intensity per unit area. Based on this formulation, the absorbed light intensity I_a is expressed as $I_a = I_0\epsilon[A]$.

The rate of polymerization is given by

$$-\frac{d[M]}{dt} = k_p[M][P_*]. \qquad (4.2.26)$$

Equations (4.2.24)-(4.2.26) were solved numerically for parameters chosen to simulate the properties of the bulk polymerization of diethylene glycol diacrylate (DEGDA) with 2,2-dimethoxy-2-phenylacetophnone as an initiator (Table 1 in Kurdikar and Peppas 1994). Figure (4.2.6) presents theoretical values from the model for the rates of propagation and termination as a function of monomer conversion or reaction time. The change in these rates is qualitatively similar to those in Figs.(4.2.3), (4.2.4), and (4.2.5) from other diffusion-controlled models. The model overpredicted the final conversion attained. This result was attributed to the fact that termination through chain transfer was not considered in the model and/or to the calculation of the monomer diffusion coefficient under the assumption of a lightly cross-linked polymerization system.

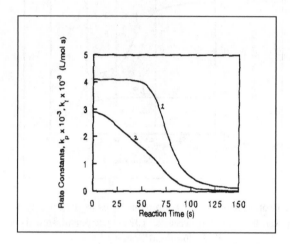

Fig. 4.2.6. Propagation (curve 1) and termination (curve 2) rate constants predicted from the model as a function of polymerization time (reprinted from Kurdikar and Peppas 1994 with permission from the American Chemical Society)

The kinetic scheme in Eq.(4.2.1) which includes initiation, propagation, and termination has been refined by Hutchinson et al. (1994) to include different termination steps as well as chain transfer to monomer or solvent and used to develop a polymerization model for determining molecular weight distribution (MWD) under pulsed-laser conditions. Radicals generated during the laser pulse grow by propagation during the dark period. Polymer radicals, not terminated during the dark period, have a high probability of termination with newly initiated radicals during the next light pulse. As a result, the molecular weight distribution (MWD) shows a sharp peak at a chain length, P_0, given by the approximate relation

$$P_0 = k_p[M]t_0, \qquad (4.2.27)$$

where $[M]$ is the monomer concentration, and t_0 is the dark period between laser pulses. Radical chains that do not terminate after one pulse from radical initiation contribute to peaks at chain lengths kP_0 ($k = 2, 3, 4, ...$). The chain length P_0 corresponds approximately to the inflection point on the low molecular weight side of the peak (Olaj et al., 1987). From Eq. (4.2.27), one may estimate the propagation rate k_p. Monte Carlo simulation by O'Driscoll and Kuindersma (1993) showed that this estimation method was accurate to within 3%. The kinetic steps used for the development of the mathematical model for pulsed-laser polymerization (PLP) is shown in the following scheme:

$$I \rightarrow 2R_*$$

$$R_* + M \xrightarrow{k_i} RM_*$$

$$RM_{n*} + M \xrightarrow{k_p} RM_{n+1*}$$

$$RM_{n*} + M \xrightarrow{k_{t,m}} D_n + RM_* , \ transfer \ to \ monomer$$

$$RM_{n*} + S \xrightarrow{k_{t,S}} D_n + RM_* , \ transfer \ to \ solvent \qquad (4.2.28)$$

$$RM_{n*} + RM_{m*} \xrightarrow{k_{t_d}} D_n + D_m, \ \text{termination by disproportionation}$$

$$RM_{n*} + RM_{m*} \xrightarrow{k_{t_c}} D_{n+m}, \ \text{termination by combination}$$

$$RM_{n*} + R_* \xrightarrow{k_t} D_n$$

$$R_* + R_* \xrightarrow{k_t} \text{terminated}; \ k_t = k_{t_c} + k_{t_d},$$

where D_n represents a dead chain of length n.

From this kinetic scheme and chain balance, one has for initiator radicals, R_*,

$$\frac{d[R_*]}{dt} = [\Delta R_*] \sum_{j=0}^{N} \delta(t - jt_0) - (k_i[M] + k_t\mu_0 + k_t[R_*])[R_*], \qquad (4.2.29)$$

$$R_*(0) = [\Delta R_*], \qquad (4.2.30)$$

where $[\Delta R_*]$ is the radical concentration generated per laser pulse, $\delta(t - jt_0)$ is the Dirac delta function, and $\mu_0 = \sum_{n=1}^{\infty} RM_{n*}$ is the total concentration of the growing radical chains.

For a chain of length n,

$$\frac{dRM_{n*}}{dt} = k_p[M]RM_{n-1*} - (k_p[M] + k_{t_m}[M]$$
$$+ k_{t_S}[S] + k_t\mu_0 + k_t[R_*])RM_{n*};$$
$$RM_{n*}(0) = 0, \ n \geq 2. \qquad (4.2.31)$$

For dead chains of length n,

$$\frac{dD_n}{dt} = (k_{t_m}[M] + k_{t_S}[S] + k_{t_d}\mu_0 + k_t[R_*])RM_{n*}$$
$$+ \frac{1}{2}k_{t_c} \sum_{m=1}^{n-1} RM_{m*}RM_{n-m*};$$
$$D_n(0) = 0, \ n \geq 2. \qquad (4.2.32)$$

In solving Eqs. (4.2.29) to (4.2.32), the maximum chain length allowed was 7500. As a result, the chain fractions for the molecular weight distributions were calculated as a number fraction

$$f(n) = \frac{D_n}{\lambda_0} \; ; \; \lambda_0 = \sum_{n=1}^{\infty}(D_n + RM_{n*}), \qquad (4.2.33)$$

or a weight fraction

$$w(n) = \frac{nD_n}{\lambda_1} \; ; \; \lambda_1 = \sum_{n=1}^{\infty} n(D_n + RM_{n*}). \qquad (4.2.34)$$

In this formulation,

$$\frac{d\lambda_0}{dt} = k_i[M][R_*] + (k_{t_m}[M] + k_{t_s}[S])\mu_0 - \frac{1}{2}k_{t_c}\mu_0^2 \; ; \; \lambda_0(0) = 0, \qquad (4.2.35)$$

$$\frac{d\lambda_1}{dt} = -\frac{d[M]}{dt} = k_i[M][R_*] + (k_{t_m}[M] + k_{t_s}[S] + k_p[M])\mu_0 \; ; \lambda_1(0) = 0, \qquad (4.2.36)$$

and

$$\frac{d\mu_0}{dt} = k_i[M][R_*] - k_t\mu_0(\mu_0 + [R_*]) \; ; \; \mu_0(0) = 0. \qquad (4.2.37)$$

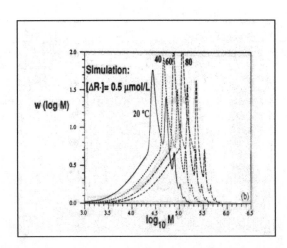

Fig. 4.2.7. Molecular weight distribution of polymethyl methacrylate (PMMA) from model simulation and experimentally from a pulsed laser at 10Hz. (reprinted from Hutchison et al. 1994 with permission from the American Chemical Society)

The above differential equations as derived from the kinetic scheme in Eq.(4.2.28) describe the dynamics of pulsed-laser polymerization. A numeric solution of the system of differential equations for certain model parameters (parameters in $L/mol\ sec$ for PMMA were $k_p = 2.39 \times 10^6 \exp(-5300/RT)$, $k_t = 1.47 \times 10^8 \exp(-701/RT)$, $k_{t_c} = 0.0$, $k_{t_m}/k_p = 7.4 \times 10^{-2} \exp(-5670/RT)$, and $k_i = 10k_p$) gave the molecular weight distributions in Fig. (4.2.7) which show similar trends to the experimental results. However, model results show sharper peaks as well as more peaks (tertiary and higher order) than the experimental curves. In the figure, $w(logM) = \log Mw(M)$. From such a model, one may have an understanding of the relative importance of model parameters on the molecular weight distribution or monomer conversion and gain insights for defining appropriate experimental conditions.

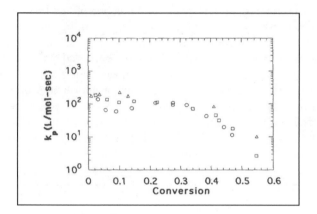

Fig. 4.2.8. Propagation kinetic constant as a function of monomer conversion for DEGDMA polymerized at 30^0C (O), 40^0C ($\square$), 50^0C ($\triangle$), and 4.7mW/cm^2 light intensity with DMPA as intiator (reprinted from Anseth et al. 1994 with permission from Elsevier Science)

Experimental work to characterize the kinetic constants and reaction behavior for photopolymerization of multi(meth)acrylate monomers was undertaken by Anseth et al. (1994). In light of the diffusion-controlled models, of interest are the changes in the propagation and termination rate constants as a function of monomer conversion. These rates were estimated from the expressions

$$\frac{k_p}{k_t^{1/2}} = \frac{R_p}{(\phi I_0 \epsilon[A])^{1/2}[M]}, \tag{4.2.38}$$

and

$$k_t^{1/2} = \frac{k_p/k_t^{1/2}}{2(t_1 - t_0)} \left(\frac{[M]_{t=t_1}}{R_p\,|_{t=t_1}} - \frac{[M]_{t=t_0}}{R_p\,|_{t=t_0}} \right), \tag{4.2.39}$$

where $t = t_0$ is the beginning of the dark period, and $t = t_1$ is a later time in the dark period. Here, $R_p = -\frac{d[M]}{dt}$ was measured as a function of time.

Derivation of these equations assumes that k_t and k_p remain constant within the small time interval $(t_1 - t_0)$ where monomer conversion is small. Equations (4.2.38) and (4.2.39) can be obtained from Eqs. (4.2.6) and (4.2.10) where k_t in Eq. (4.2.10) is replaced by 2 k_t. Figures (4.2.8) and (4.2.9) show the propagation and termination rates as a function of monomer conversion for DEGDMA polymerized at $30^0 C$ (O), $40^0 C$ (□), $50^0 C$ (△), and $4.7 mW/cm^2$ light intensity with DMPA as initiator.. These experimental observations are in qualitative agreements with predicted results for the diffusion- controlled models in Figs. (4.2.3) to (4.2.6). It is seen from Fig.(4.2.8) that the propagation rate is approximately constant until 40% conversion after which it becomes diffusion-controlled and begins to decrease with an increase in conversion. It is interesting to note that the behavior of the termination rate is similar to that in Fig (4.2.4) as predicted from the diffusion-controlled model of Anseth and Bowman (1993). At low conversion, termination is diffusion-controlled and decreases with time. Between 20% and 40% conversion termination becomes approximately constant which can be explained as being the result of reaction diffusion control. At higher conversion, propagation becomes diffusion-limited, and termination decreases once more. There is a slight increase in the propagation and termination rates as temperature increases. Similar results were observed for TrMPTrMA (a trimethacrylate monomer) and DEGDA as a function of conversion.

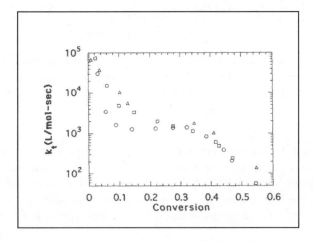

Fig. 4.2.9. Termination kinetic constant as a function of monomer conversion for DEGDMA polymerized at 30^0C(O), 40^0C(□),50^0C(△), and 4.7mW/cm^2 light intensity with DMPA as intiator (reprinted from Anseth et al. 1994 with permission from Elsevier Science)

However, no significant plateau was observed for propagation or termination in the case of other monomers (TrMPTrA, PETeA, and DPEMHPeA).

This result was explained by an increase in viscosity which is caused by higher molecular weight as a result of the higher functionality of these polymers.

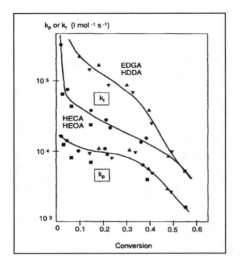

Fig. 4.2.10. Propagation (k_p) and termination (k_t) constants as a function of monomer conversion for polyurethane-acrylate containing different types of monomers in a 1:1 weight ratio. EDGA (▼), HDDA (▲), HECA (●), HEOA (■). I=20mW/cm^2 (reprinted from Decker et al. 1996, p 181, by courtesy of Marcel Dekker Inc.)

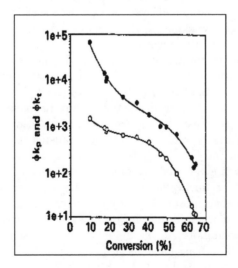

Fig. 4.2.11. Changes of ϕk_p and ϕk_t as a function of percent conversion (reprinted from Lecamp et al. 1999 with permission from Elsevier Science)

Variation of k_p and k_t with monomer conversion, similar to the above results of Anseth et al.(1994), were observed for polyurethane-acrylate and dimethacrylate as seen in Fig.(4.2.10) from Decker et al. (1996) and in Fig.(4.2.11) from Lecamp et al. (1999). In Fig. (4.2.10), polyurethane-acrylate was used with different monomers in a 1:1 weight ratio. The monomers were ethyldiethyleneglycol monoacrylate (EDGA), hexanediol diacrylate (HDDA), hydroxyethyl-oxazolidone monoacrylate (HEOA), and hydroxyethyl-carbamate monoacrylate (HECA). Dimethoxyphenylacetophenone (DMPA) was used as initiator at 3% concentration. Real-time infrared spectroscopy was used to measure monomer conversion (or polymerization rate) as a function of exposure time from which variations in the propagation and termination rates over time were estimated using Eqs (4.2.6) and (4.2.10) with k_t replaced by $2\ k_t$. In Fig.(4.2.11) dimethacrylate oligomer was used with 2,2-dimethyl-2-hydroxyacetophenone as initiator. Estimates of ϕk_p and ϕk_t were obtained using essentially Eqs. (4.2.38) and (4.2.39) of the kinetic model by Tryson and Shultz (1979).

4.3 Photofabrication

In laser photopolymerization, prototypes of various shapes and complexity can be fabricated by controlling the light exposure of the monomer resin. This feature enables the method to be used effectively in the manufacturing of components for assembly or for joining together different microparts to form complicated structures. In photoforming, light exposure for generating the shape of the microobject can be applied through a mask or by free-form scanning of the monomer layer. Also, a free surface or a fixed surface method can be used for exposing the monomer in the vat to be solidified. Figure (4.3.1) is a schematic diagram of the free surface method and the fixed surface method. In the former, it is seen that the monomer layer is exposed to the beam on top of the elevator platform and the platform is lowered layer by layer through the resin causing the microobject to be manufactured (by stacking the solidified layers) from the bottom up. In the latter, the monomer layer is exposed to the beam on the window surface below the elevator platform and the platform is raised layer by layer causing the microobject to be manufactured from the top down. In the free surface method, there is less control over the evenness of the monomer surface which can lead to a higher chance of deformation during solidification. The scanning method gives better control of the light exposure point by point than the mask exposure. This control can lead to more uniformity or evenness in the degree of solidification which minimizes nonuniform shrinkage of the resin. Based on the above considerations, Takagi and Nakajima (1993, 1994) used the fixed surface method with scanning in the set-up depicted in Fig.(4.3.2) to study the resolution of this microfabrication process.

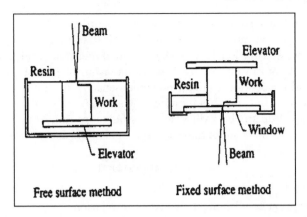

Fig. 4.3.1. Schematic diagram of a free surface and a fixed surface photofabrication method (Takagi and Nakajima 1993 with permission from IEEE)

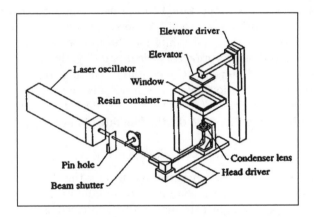

Fig. 4.3.2. Scematic diagram of a photofabrication apparatus (Takagi and Nakajima 1993 with permission from IEEE)

Achieving high resolution or precision is a major issue in photofabrication. For improving precision, it is essential to control the shape and dimensions of the solidified volume caused by the spotlight of the beam during scanning of the monomer layer on top of the window surface. For optimum results, the solidified volume or cell should have nearly the same dimensions as those dictated by the x, y scanning movement of the beam and by the z-movement of the platform. As such, isometric cell shapes are desirable for equal displacements in the x, y, and z directions commonly used in practice. Solidified cell dimensions and shape are influenced by conditions of light exposure and by the characteristics of the monomer. Transmission, scattering (as a Brownian movement), and absorption behavior of light photons in the monomer influences the solidification process and precision. Scattering in the x,y directions contributes to deformation in cell shape and dimensions and hence to less precision in photoforming. Monte Carlo simulation of photon transmission, scattering, and absorption to determine the density distribution of absorbed photons indicated that, for thin monomer layers that are solidified in practice, photon transmission in the z-direction was far more important than photon scattering. Since scattering is minimal, the light intensity at depth z may be expressed as

$$I_z = I_0 \exp(-\alpha z), \tag{4.3.1}$$

where I_z is the light intensity at z, I_0 is the incident light on the surface, and α is the absorption coefficient. This gives the relation

$$\frac{\Delta I_z}{I_z} = 1 - \exp(-\alpha \Delta z) \doteq \alpha \Delta z. \tag{4.3.2}$$

For small Δz, Eq.(4.3.2) indicates that the depth of photon transmission and therefore solidification is proportional to the light amount and is in inverse relation to the absorption coefficient. A more fundamental approach based on the kinetics of initiation and inhibition shows that the depth of cure is inversely related to the absorption coefficient and directly proportional to the logarithm of the light intensity (Cook, 1991). The occurrence of a specific depth of cure or solidification below which no polymerization occurs can be explained by the consumption of free radicals by inhibitor species and by the attenuation of light intensity with depth. The rate of initiation, R_i, at a depth z is given by

$$R_i = \phi \epsilon_s [I] I_0 \exp(-\alpha z), \tag{4.3.3}$$

where $[I]$ is the photoinitiator concentration and ϵ_s is the absorption coefficient of the photo-initiator. The local concentration of inhibitor, $[x]$, at a particular depth after a period t of light exposure can be expressed as

$$[x] = [x_0] - R_i t, \tag{4.3.4}$$

where $[x_0]$ is the initial concentration of the inhibitor. Since the depth of cure occurs where the inhibitor concentration is depleted ($[x] \approx 0$), the cure depth, as seen from Eqs. (4.3.3) and (4.3.4), is

$$z = (\frac{1}{\alpha}) \ln(\phi \epsilon_s [I] I_0 t / [x_0]. \tag{4.3.5}$$

Expression (4.3.5) shows that, in addition to α and I_0, cure depth is also a function of the initial concentrations of initiator and inhibitor as well as the absorption coefficient of the photo-initiator and exposure time. However, Eq. (4.3.2) is a simplified approach that gives the correct guidelines as to important photoforming factors (namely, light absorption and monomer resin characteristics) to optimize for precision control. Soft focusing of the laser was found to be an easy way to control the light intensity of the beam. Soft focusing to make the diameter of the exposure spot the same as that of the scanning rule (or size of the cell in the x,y directions) gave better cell formation than sharp focusing. The diameter and stability of exposure light have an effect on the vertical resolution of the microobject. Likewise, high absorption characteristic of the resin leads to better vertical resolution. Figures (4.3.3) and (4.3.4) show solidified cells for very sharp focusing ($D = 0.0\ mm$) and for soft focusing ($D = 0.16mm$). Here, D is the offset between the focal point of the condenser lens and the surface of the monomer resin. A D value of 0.16 mm translates into a light exposure spot 60 μm in diameter.

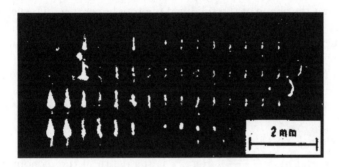

Fig. 4.3.3. Solidified cells using sharp focusing (Takagi and Nakajima 1993 with permission from IEEE)

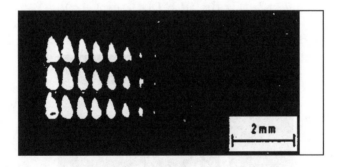

Fig. 4.3.4. Solidified cells using soft focusing (Takagi and Nakajima 1993 with permission from IEEE)

The cells in Fig.(4.3.3) are 500 μm long in the vertical direction and 40 μm in diameter. On the other hand, with soft focusing [Fig.(4.3.4)] the length of a cell is 60 μm with a diameter of 40 μm. Soft focusing is in better agreement with the theoretical argument leading to better resolution when the cells are put together in forming a microobject. Using this approach, a resolution within 10 μm was achieved. Figure (4.3.5) shows a photofabricated micro-clamping tool. The clamp can be made to open and close by an actuator positioned below the substrate of the clamping device and connected to it by a steel pin. When voltage is applied to the actuator, the pin is raised causing the clamping arms to open. The resolution of this device was 100 μm vertically and 20 μm horizontally.

The accuracy of high aspect ratio microstructures produced by photofabrication using different methods of exposure (pattern transfer of a uniform beam through a mask, direct writing with a shaped beam of uniform intensity, and direct writing with a focused Gaussian beam) has been examined by simulating the shape of the solidified photopolymer using Fresnel's diffraction theory in conjunction with light absorption (Yamaguchi et al. 1994a,b). When using a mask (with a square opening of dimension l) for forming a solid pattern on the polymer surface ($z = 0$) , the light intensity at (x, z) after diffraction (Sommerfeld, 1954; Yamaguchi et al., 1994a) may be expressed as

$$I_d(x, z) = \frac{1}{2}I_0\{[C(s_2) - C(s_1)]^2 + [S(s_2) - S(s_1)]^2\}, \qquad (4.3.6)$$

where $C(s) = \int_0^s \cos(\frac{\pi u^2}{2})du$, $S(s) = \int_0^s \sin(\frac{\pi u^2}{2})du$, $s_1 = \frac{2}{\sqrt{m}}(\frac{x}{l} - \frac{1}{2})$, $s_2 = \frac{2}{\sqrt{m}}(\frac{x}{l} + \frac{1}{2})$, and $m = \frac{2\lambda(h+z)}{l^2}$. Here, z is the depth of the monomer resin, λ is the laser light wave length, and h is the distance between the mask and the surface of the resin. The light intensity per unit area at depth z for an exposure time t is given by

$$E(x, z, t) = I(x, z)t = I_d(x, z)\exp(-\alpha z)t, \qquad (4.3.7)$$

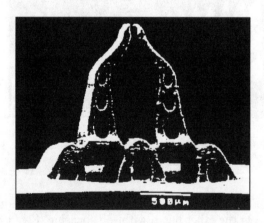

Fig. 4.3.5. Photofabricated micro-clamping tool (Takagi and Nakajima 1994 with permission from IEEE)

If the polymer is assumed to solidify at an energy threshold $E_0 = E(x, z, t)$, then

$$I(x, z) = \frac{E_0}{t}.$$
(4.3.8)

The line (x, z) satisfying Eq.(4.3.8) gives the shape of the cross-section of the solidified polymer. Evaluation of Eq.(4.3.6) assuming a straight edge in the z direction at $x = 0$ and l infinity (Sommerfeld, 1954) gives

$$\frac{I_d(x, z)}{I_0} = \frac{1}{2}\left\{\frac{1}{2} + C\left[x\sqrt{\frac{2}{\lambda(h + z)}}\right]\right\}^2 + \frac{1}{2}\left\{\frac{1}{2} + S\left[x\sqrt{\frac{2}{\lambda(h + z)}}\right]\right\}^2.$$
(4.3.9)

For ease of computation, Eq. (4.3.9) was approximated by the empirical expression

$$\frac{I_d(x, z)}{I_0} = a\exp(bx\sqrt{\frac{2}{\lambda(h + z)}}), \quad a = 0.2486, \text{ and } b = 1.649.$$
(4.3.10)

It was shown that Eq.(4.3.10) approximates the exact solution in Eq.(4.3.9) for $x\sqrt{\frac{2}{\lambda(h+z)}} \leq 1.0$.

Substituting Eqs. (4.3.10) and (4.3.8) into Eq. (4.3.7) gives x, the outline of the formed solid,

$$x = \frac{1}{b}\sqrt{\frac{\lambda(h + z)}{2}}\left[\alpha z - \ln(\frac{I_0 a t}{E_0})\right].$$
(4.3.11)

The value for x as a function of z gives the predicted deviation of the sidewall from a straight edge ($x = 0$). Since one edge of the mask opening

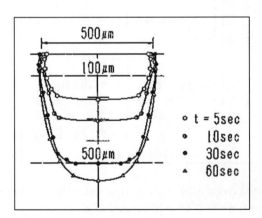

Fig. 4.3.6. Width and depth of a solidified photopolymer as a function of exposure time, t (reprinted from Yamaguchi et al. 1994a with permission from ASME)

is at $x = 0$ and the other at $x = l$, a negative x value indicates an increase in width of the solidified polymer. Considering the change in width at both edges in the cross-section, an increase in width with depth may be set equal to $-2x$. Hence,

$$\delta = -2x = \frac{\sqrt{2\lambda(h+z)}}{b}\left[\ln(\frac{I_0 a t}{E_0}) - \alpha z\right]. \qquad (4.3.12)$$

As can be seen from Eq.(4.3.12), the deviation from a straight wall is reduced by decreasing the distance, h, between the mask and the polymer surface. Also, δ increases with an increase in exposure time or depth of the solidified photopolymer. Figure (4.3.6) shows experimental observations on width and depth of the cross-section of a solidified photopolymer. As predicted from Eq. (4.3.12), width increased with an increase in exposure time or depth of the solid. Figure (4.3.7), for $h = 1000\ \mu m$, shows fairly good agreement between observed width and the predicted width from Eq. (4.3.12). The accuracy in width at $z = 100\ \mu m$ and for $t = 6\ secs$ was reported to be $\pm\ 2$ percent. From Eq. (4.3.12), the accuracy for $t = 6$ secs and $h = 1000 \mu m$ for the same depth is calculated to be ± 6 percent. An accuracy of $\pm\ 1$ percent is predicted for a smaller h of $10\ \mu m$. Since accuracy in the sidewall decreases with depth, one would expect high accuracy if the microstructure is built layer by layer where the thickness of each layer is small.

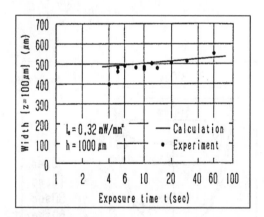

Fig. 4.3.7. Predicted (solid line) and experimental (dots) values for the width of a solidified polymer (reprinted from Yamaguhi et al. 1994a with permission from ASME)

For direct writing by a circular beam of radius r_c and of uniform intensity, the light intensity after diffraction (Linfoot, 1955; Yamaguchi et al., 1994b) can be expressed as

$$I_d(r,z) = \frac{4\pi^2 I_0}{\lambda^2(h+z)^2} \Bigg[\left(\int_0^{r_c} \cos(\frac{\pi\rho^2}{\lambda(h+z)}) J_0(\frac{2\pi\rho r}{\lambda(h+z)}) \rho d\rho \right)^2$$

$$+ \left(\int_0^{r_c} \sin(\frac{\pi\rho^2}{\lambda(h+z)}) J_0(\frac{2\pi\rho r}{\lambda(h+z)}) \rho d\rho \right)^2, \qquad (4.3.13)$$

where J_0 is the 0 order type 1 Bessel function. Calculations based on Eq. (4.3.13) and Eqs. (4.3.7) and (4.3.8), with x replaced by r, showed that the sidewalls were tapered inward with depth. Also, the width and depth of the solidified polymer increased with an increase in irradiation time. These results are similar to what is observed in Figs. (4.3.6) and (4.3.7) for the case of a square pattern and with irradiation of the surface through a mask.

When the monomer surface is exposed to a Gaussian beam of radius w, the light intensity is expressed as

$$I_d(x,y,z) = \frac{2I_0}{\pi w^2} \exp(-\frac{2(x^2+y^2)}{w^2}). \qquad (4.3.14)$$

The intensity of the beam, $I(x,y,z)$, at depth z is $I(x,y,z) = I_d(x,y,z)e^{-\alpha z}$.

When the beam is scanned on a straight line along the y-axis at a constant velocity v (direct shaped beam writing), the energy per unit area at (x,y,z) is given by

$$E(x,y,z) = \int_{-\infty}^{\infty} I(x,y-vt,z)dt. \qquad (4.3.15)$$

Assuming that the polymer solidifies at a threshold energy value E_0, the line (x,z) satisfying Eq.(4.3.15) forms an outline of the shape of the solid. Theoretical calculations by Yamaguchi et al. (1994b) based on Eq. (4.3.15) showed that at a scan speed of 25 $\mu m/sec$, the difference between the width of the solidified polymer at the surface and at a depth of 200 μm was 4 μm, 6 μm, and 135 μm for a square, circular, and Gaussian beam, respectively. Thus, direct writing using a Gaussian beam is least accurate when sidewalls are considered.

Considering direct writing using a focused beam (by passing through a lens), the beam intensity can be expressed as

$$I_d(x,y,z) = \frac{2I_0}{\pi w^2(z)} \exp(-\frac{2(x^2+y^2)}{w^2(z)}), \qquad (4.3.16)$$

When the beam focal point is at the surface of the monomer liquid, its radius changes with depth and is given by

$$w(z) = w_{min}\sqrt{1 + \left(\frac{\lambda z}{\pi w_{min}^2}\right)^2}, \qquad (4.3.17)$$

where $w_{min} = \frac{f\lambda}{\pi R}$ and represents the minimum beam radius on the surface of the monomer resin. Here, f is the focal length of the lens, R is the beam radius entering the lens, and λ is the wavelength. When the beam is scanned with a constant velocity v on a straight line along the y-axis, the half-width of the formed solid is given by

$$x = \frac{w(z)}{\sqrt{2}} \sqrt{\ln\left(\sqrt{\frac{2}{\pi}} \frac{I_0}{E_0 w(z) v}\right) - \alpha z}. \qquad (4.3.18)$$

Calculations by Yamaguchi et al. (1994b) showed that at a scan speed of 40 $\mu m/sec$ a solid can be formed with a difference in width (measured at the surface and at a depth of 100 μm) of 0.2 μm. Thus, by selecting the optimum scan speed for a focused Gaussian beam, microstructures can be formed with higher sidewall accuracy than that attained by using a mask or by direct writing using a square or a circular beam.

Optimum conditions for producing high aspect ratio microstructures when the beam focal point is at the surface were examined using Eq. (4.3.17) and (4.3.18). The solid boundary $(x/R, z/R)$ was determined by varying the parameters $\frac{\lambda}{R}, \frac{f}{R}, \frac{\alpha}{R}$, and $\frac{P}{E_0 R v}$. Each parameter was varied, holding the others constant, in order to determine its effect on the aspect ratio, defined for a constant depth L as $\frac{L}{2x(\max)} \frac{x(\min)}{x(\max)}$ (Yamaguchi et al., 1995; Nakamoto et al.,1996). For a given $x(\max)$, the maximum occurs when the wall is straight or $\frac{x(\min)}{x(\max)} = 1$. As the wave length increased, it was found that the ratio increased to a maximum and then decreased. The maximum occurred at $\lambda = 0.15$ μm for $L = 50\mu m$, $I = 0.1\mu W$, $\frac{f}{R} = 20$, $\alpha = 6.94 x 10^{-3} \mu m^{-1}$, and $E_0 = 0.34$ mJ mm^{-2}. Likewise, the optimum, when varying $\frac{f}{R}$ for $\lambda = 0.325$ μm with the same values for the other parameters, was $\frac{f}{R} = 18$. Varying the absorption coefficient α, the aspect ratio reached its maximum value when α was zero.

Optimum conditions were examined also when the focal point was not set at the surface (defocusing). The starting point in this analysis is the approximate expression

$$w(z) = w_s \pm z\frac{R}{f}, \qquad (4.3.19)$$

where w_s is the beam radius at the surface of the monomer liquid and $z\frac{R}{f}$ is the approximate solution to Eq. (4.3.17) when $z \gg \frac{\lambda f^2}{R^2}$. In Eq. (4.3.19), the focal point is above the surface of the monomer resin when the sign is plus and below when minus. A desired optimum is to have the walls of the solidified photopolymer perpendicular to the surface. This condition is attained by substituting Eq. (4.3.19) into Eq. (4.3.18), calculating the first and second partial derivatives of x with regard to z in the resulting equation, and setting the derivatives (evaluated at $z = 0$) equal to zero. This gives

$$w(z) = w_s - \frac{3}{2}\alpha w_s z, \qquad (4.3.20)$$

and

$$\ln\left(\sqrt{\frac{2}{\pi}}\frac{I_0}{E_0 w_s v}\right) = \frac{1}{6}. \tag{4.3.21}$$

From Eqs. (4.3.19), (4.3.20) and the relation $w(0) = \frac{f\lambda}{\pi R}$, one has

$$w_s = \frac{2R}{3\alpha f} = \frac{2\lambda}{3\pi\alpha w_{\min}}. \tag{4.3.22}$$

When the focal point is below the surface of the polymer at z_0, Eq. (4.3.17) becomes

$$w(z) = w_{\min}\sqrt{1 + \left(\frac{\lambda(z - z_0)}{\pi w_{\min}^2}\right)^2}. \tag{4.3.23}$$

Replacing $w(z)$ by w_s for $z = 0$ in Eq.(4.3.23) and solving for z_0, one obtains

$$z_0 = \frac{\pi w_{\min}}{\lambda}\sqrt{w_s^2 - w_{\min}^2}. \tag{4.3.24}$$

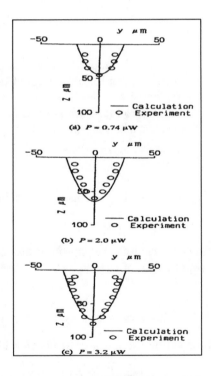

Fig. 4.3.8. Predicted and observed profile of the solidified polymer when the focal point is set at the monomer surface (λ=0.325μm, $\frac{f}{R} = 192$, v=10 μm s^{-1}, $\alpha = 4.9\times10^{-2}\mu$m^{-1}, $E_0 = 0.34$mJ mm^{-2}) (reprinted from Nakamoto et al. 1996 with permission from IOP Publishing Limited)

To obtain the solidified depth for the case where the wall is perpendicular to the surface, one must determine w_{min} from the relation $w_m = \frac{f\lambda}{\pi R}$ after which w_s and z_0 are calculated from Eqs. (4.3.22) and (4.3.24). with w_s known, $\frac{I_0}{v}$ is calculated from Eq. (4.3.21). For application purposes, either I_0 or v must be specified. The solidified polymer obtained when the focal point was below the surface showed a smaller width and a higher aspect ratio than that obtained when the focal point was at the surface.

Comparisons with experimental data, Figure (4.3.8), shows good agreement between observed and predicted solid profile from Eq.(4.3.18) when the focal point was at the surface. The difference in width or depth between predicted and observed was within ± 5 μm. For the experiment, a He-Cd laser was used with $\lambda = 0.325$ μm, and $R = 3.6mm$. The monomer was an unsaturated polyester (APR stamp resin, product of Asahi Chemical Co.). Rhodamine B ($C_{28}H_{31}ClN_2O_3$) was added to the monomer resin in order to increase its absorption coefficient. Parameters used were $\lambda = 0.325$ μm, $\frac{f}{R} = 192$, $v = 10$ μm s^{-1}, $\alpha = 4.9 \times 10^{-2}$ μm^{-1}, $E_0 = 0.34$ mJ mm^{-2} with $I_0 = 0.74, 2.0$, and 3.2 μW for (a), (b), and (c), respectively.

4.4 Process Models

A model to account for spacial and temporal variations in monomer conversion, photo-initiator concentration, and heat generated by photopolymerization in a small cylindrical region or volume element exposed to the laser beam was considered by Flach and Chartoff (1995a). For a Gaussian laser beam, the incident light intensity on the surface was expressed as

$$I(r, z, t) = I_0 \exp(-\frac{2r^2}{w_0^2}); \ r \geq 0, \ z = 0, \ t \geq 0, \tag{4.4.1}$$

where w_0 is the beam radius. The light intensity at a depth z is given by

$$I(r, z, t) = I(r, 0, t) \exp(-\epsilon S z), \tag{4.4.2}$$

where ϵ is the absorption coefficient of the photo-initiator and S is the initiator concentration . As such, the rate of change in light intensity with z is given by

$$\frac{\partial I(r, z, t)}{\partial z} = -\epsilon S I \tag{4.4.3}$$

The light intensity absorbed by the initiator at depth z is

$$I_a(r, z, t) = \epsilon S I. \tag{4.4.4}$$

The rate of change in S with time is

$$\frac{\partial S}{\partial t} = -\phi I_a(r, z, t), \tag{4.4.5}$$

where $S(r, z, 0) = S_0$; $r \geq 0$, $z \leq D$ = cylinder depth. Also, the rate of monomer conversion from Eq.(4.2.5) is

$$\frac{\partial [M]}{\partial t} = -k_p[M] \left[\frac{\phi I_a(r, z, t)}{k_t}\right]^{1/2}. \tag{4.4.6}$$

The temperature distribution generated by the polymerization reaction can be derived from the standard transient heat conduction equation in the cylindrical domain

$$\rho C_p \frac{\partial T}{\partial t} = k \left[\frac{1}{r}\frac{\partial}{\partial r}(r\frac{\partial T}{\partial r}) + \frac{\partial^2 T}{\partial z^2}\right] + H_p R_p, \tag{4.4.7}$$

where $R_p = -\frac{\partial [M]}{\partial t}$ is the polymerization rate and H_p is the heat generated (J/mol). The initial and boundary conditions for Eq.(4.4.7) are given by

$$T(r, z, t) = T_0; \ t = 0, \ r \geq 0, \ z \geq 0,$$

$$\frac{\partial T}{\partial r} = 0; \ r = 0, \ z \geq 0, \ t \geq 0,$$

$$T = T_0; \ r \to \infty, \ z \geq 0, \ t \geq 0, \tag{4.4.8}$$

$$T = T_0; \ z \to \infty, \ r \geq 0, \ t \geq 0,$$

and

$$\frac{\partial T}{\partial z} = 0; \ z = 0, \ r \geq 0, \ t \geq 0.$$

Equations (4.4.3)-(4.4.7) were solved simultaneously and iteratively by calculating new values for these variables at the end of every time step Δt. This involves calculating first the light intensity at z from Eqs. (4.4.1) and (4.4.2). The next step is to calculate S and M, for a time step Δt, from Eqs. (4.4.5) and (4.4.6) and determine the temperature distribution from Eqs. (4.4.7) and (4.4.8). The process is repeated for a new time step. The fourth order Runge-Kutta numerical technique was used for solving Eqs. (4.4.4)-(4.4.6) and the alternating-direction implicit method (ADI) for solving the heat conduction equation.

The model was used to simulate the photopolymerization reaction of HDDA in a cylindrical region with a diameter of $0.25mm$ and a height of 0.5 mm. The laser (15 mW HeCd 325 nm) considered had a beam radius of $1.25 \times 10^{-2}cm$ and $I_0 = 4.8 \times 10^{-5}$ E cm^{-2} sec. The ratio $\frac{k_p}{k_t^{1/2}}$ was expressed as a function of the temperature, T, and monomer conversion through the Arrhenius equation

$$\frac{k_p}{k_t^{1/2}} = k_0 \exp(-\frac{R_a}{RT}), \tag{4.4.9}$$

where k_0 and R_a were allowed to vary with monomer conversion through simple exponential functions. Other parameter values used in the simulation were the following: $S_0 = 0.15 \; mol \; l^{-1}$, $[M_0] = 3.0 \; mol \; l^{-1}$, $T_0 = 35^0 C$, $\epsilon = 200$ $mol^{-1} cm^{-1}$, $\phi = 0.1$, $\rho = 1.15 \; g \; cm^{-3}$, $k = 1.7 \; J \; g^{-1} \; K^{-1}$, and $H_p = 1.65 \times 10^{-5} J \; mol^{-1}$. The kinetic parameters for HDDA were taken from Tryson and Schultz (1979).

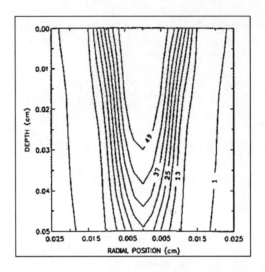

Fig. 4.4.1. Contours of percent monomer conversion as a function of depth and radial position (reprinted from Flach and Chartoff 1995 with permission from the Society of Plastics Engineers)

Salient results indicated, as expected, that monomer conversion decreased with depth and radial distance from the origin and increased with time of exposure. The temperature rise generated by the polymerization reaction increased initially up to a maximum of $35^0 C$ and then decreased as the reaction slowed down. Maximum temperature rise was at $z = 0$ and $r = 0$ and generally decreased with depth and with radial distance from the origin. Contours of monomer conversion at $300ms$ exposure time in Fig. (4.4.1) show a conical form indicating the shape of a solidified photopolymer formed by exposure to a fixed laser beam for a relatively short time period. Another result useful in practice is the effect of the initial photo-initiator concentration on percent monomer conversion. Figure (4.4.2) shows that, at low initial concentration of the photoinitiator, the percent monomer conversion is constant with depth. However, at higher concentrations, monomer conversion decreases with depth because of decreasing light intensity caused by higher absorptivity.

The same model was extended to include beam scanning. The exposed region considered was rectangular in shape with the beam moving on a straight line in the y- or x- direction across the surface (Flach and Chartoff, 1995b). The incident laser intensity on the surface is now expressed as

$$I(x,y,z,t) = I_0 \exp\{-2[(x-x_0)^2 + (y-y_0)^2]/w_0^2\};$$
$$0 \le x \le X, \ 0 \le y \le Y, \ 0 \le z \le Z. \tag{4.4.10}$$

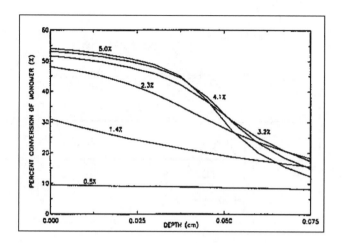

Fig. 4.4.2. Percent monomer conversion as a function of depth for different initial photoinitiator concentrations (reprinted from Flach and Chartoff 1995 with permission from the Society of Plastics Engineers)

In Eq. (4.4.10), (x_0, y_0) represents the center of the laser beam. Scanning was along a straight line. Equations (4.4.2)-(4.4.6) remain unchanged except for the coordinate system change from (r, z) to (x, y, z). Equation (4.4.7) becomes

$$\rho C_p \frac{\partial T}{\partial t} = k\left(\frac{\partial^2 x}{\partial x^2} + \frac{\partial^2 y}{\partial y^2} + \frac{\partial^2 z}{\partial z^2}\right) + H_p R_p. \tag{4.4.11}$$

The initial and boundary conditions for Eq. (4.4.11) were given by

$$T = T(x,y,z,t) = T_0; \ t = 0$$

$$\frac{\partial T}{\partial x} = 0; \ x = 0, \ t \ge 0,$$

$$T = T_0; \ x = X, \ t \ge 0,$$

$$T = T_0; \ y = 0, \ t \ge 0, \tag{4.4.12}$$

$$T = T_0; \ y = Y, \ t \ge 0,$$

$$\frac{\partial T}{\partial z} = 0; \ z = 0, \ t \ge 0,$$

and

$$T = T_0; \ z = Z, \ t \ge 0.$$

If polymerization during the dark period between laser pulses is to be included in the model, then Eq. (4.4.6) can be re-derived for the dark and light periods. The solution is given in Eq. (4.2.4) with $P_*(t)$ given in Eqs. (4.4.14) and (4.4.15).

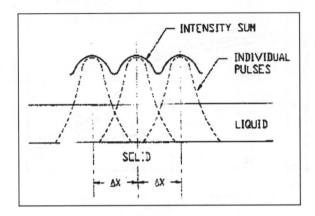

Fig. 4.4.3. Side and top view illustrations of the cusps that can form during scanning of the surface by a laser beam with a Gaussian intensity distribution (reprinted from Nassar et al. 1999 with permission from SPIE)

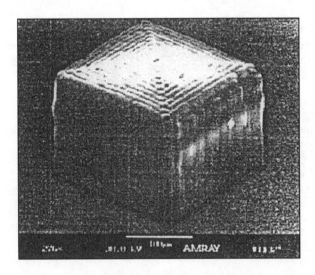

Fig. 4.4.4. A rectangular solid, formed by laser-induced polymerization, showing inaccuracies in the side walls and corrugation in the surface finish (reprinted from Nassar et al 1999 with permission from SPIE)

A numerical solution was obtained for the same conditions of the fixed beam case in Eqs.(4.4.4)-(4.4.7). The laser was assumed to move in steps with a step size of $0.00125cm$ and a step period of $.025sec$. Salient results included the temperature profile along the scan line and cure depth and width as a function of the scan rate. Predictions of monomer conversion profiles as a function of scan rate and light intensity can be obtained from such a model which are useful in process operation.

For the fabrication of three-dimensional microstructures of a pre-specified geometry with smooth surface finish and dimensional accuracy, it is necessary to determine the laser scan pattern on the surface in the x,y directions. Figure (4.4.3) is a side view illustration of the general problem that relates monomer conversion, which affects surface finish and dimensional accuracy of the photo-formed microstructure, to the irradiation of a monomer by a pulsed laser. Because of the Gaussian beam distribution, corrugations can occur in the solid when scanning the surface. These corrugations are due to inhomogeneous monomer conversion which can lead to surface roughness and imprecise control of the side walls as seen in the rectangular solid in Fig. (4.4.4). A simple modeling approach for determining the laser scan pattern in the form of pixel dwell times on the monomer surface for attaining a uniform monomer conversion by which accuracy in solid formation can be achieved has been developed by Nassar et al. (1999). The model considers polymerization as a function of time during the light and dark periods of the pulsed laser. The kinetics of free-radical polymerization is described in Eq. (4.2.1). We consider first bimolecular termination and express, from Eq. (4.2.2), the rate of change of radical concentration as

$$\frac{d[P_*]}{dt} = \Phi I_a - k_t[P_*]^2. \tag{4.4.13}$$

A solution of Eq. (4.4.13) over the time duration of the pulse or light period $(0 \le t \le t_0)$ with initial condition $[P_*]_0$ gives

$$[P_*(t)] = \sqrt{\frac{\phi I_a}{k_t}} \left(\frac{1 - e^{-2qt}}{1 + e^{-2qt}} \right), \tag{4.4.14}$$

where $q = (\phi I_a/k_t)^{1/2}$.

During the dark period or time between pulses $(t_0 \le t \le t_1)$, the solution to Eq. (4.4.13), with $\phi = 0$ and initial condition $[P_*(t_0)]$ from Eq. (4.4.14), gives

$$[P_*(t)] = \frac{[P_*(t_0)]}{[P_*(t_0)]k_t(t - t_0) + 1}. \tag{4.4.15}$$

The rate of monomer conversion during a laser pulse is

$$\frac{d[M]}{dt} = -k_p[P_*(t)][M] \tag{4.4.16}$$

with initial condition $[M]_0$ at $t = 0$.

A solution to Eq. (4.4.16) gives

$$\ln \frac{[M]}{[M]_0} = -k_p \int_0^{t_1} [P_*(\tau)]d\tau. \tag{4.4.17}$$

The integrand in Eq. (4.4.17), $[P_*(t)]$, is given by Eq. (4.4.14) in the interval $(0 \leq \tau \leq t_0)$ and by Eq. (4.4.15) in the interval $(t_0 \leq \tau \leq t_1)$. Integrating the right hand side of Eq. (4.4.17) yields

$$\ln \frac{[M]}{[M]_0} = -k_p \ln \left(\frac{\exp(qt_0) + \exp(-qt_0)}{2} \right) - \frac{k_p}{k_t} \ln \left(k_t[P_*(t_0)](t_1 - t_0) + 1 \right), \tag{4.4.18}$$

or

$$\frac{[M]}{[M]_0} = \frac{1}{\left(\frac{\exp(qt_0)+\exp(-qt_0)}{2} \right)^{k_p}} \frac{1}{(k_t[P_*(t_0)](t_1 - t_0) + 1)^{k_p/k_t}}. \tag{4.4.19}$$

To keep the model simple, we have assumed constant k_p and k_t values and that monomer conversion levels off during the dark period before the start of the next pulse. This monomer conversion assumption is reasonable since the light period is quite short as compared to the dark period. The relative lengths of the light and dark periods can be set by adjusting the pulse repetition rate. As far as the polymerization and termination rates are concerned, it is known that k_p and k_t are functions of percent monomer conversion (Decker et al., 1996; Hoyle et al., 1991; Anseth et al., 1994) or polymerization time. As such, k_p and k_t can be regarded as average values over the time span of the light and dark period. A more accurate representation would be to determine from experimental observations empirical functions relating k_p and k_t to time. Based on these functions, one may subdivide the light and dark time intervals into subintervals where k_p and k_t are near constant within a subinterval, but vary between subintervals, and evaluate Eq. (4.4.17) over successive subintervals.

Equation (4.4.19) gives the proportional reduction in monomer concentration in one pulse (light and dark period) at a point (x_i, y_j, z) where (x_i, y_j) is the pixel address on the scanned surface. For scanning, one may subdivide the monomer surface into $n_1 \times n_2$ pixels where the midpoint of a pixel coincides with the laser pulse as the laser beam moves from one pixel to another. The pixel size can be determined from the scan rate and the pulse repetition rate. Here, we consider a very thin monomer layer to be solidified. For a Gaussian beam intensity profile, it is seen that the absorbed light intensity at (x_i, y_j, z) is given by

$$I_a(x_i, y_j, z) = I_a$$
$$= I_0(x_i, y_j) \frac{1}{2\pi\sigma^2} \exp \left[-\frac{[(x_i - x_k)^2 + (y_j - y_l)^2]}{2\sigma^2} \right]$$
$$\cdot \left(1 - e^{-\alpha z} \right) \Delta x \Delta y. \tag{4.4.20}$$

Because of the Gaussian distribution, it is seen that a pixel at (x_i, y_j) receives light intensity from another (x_k, y_l) pixel where the beam is focused. From Eq. (4.4.18) and considering the overlap in beam intensity from Eq. (4.4.20), it is seen that

$$
\ln \frac{[M(x_i, y_j, z)]}{[M]_0} = \sum_{k=1}^{n_1} \sum_{l=1}^{n_2} -k_p \ln \left(\frac{e^{qt_0} + e^{-qt_0}}{2} \right)
$$
$$
- \frac{k_p}{k_t} \ln \left(k_t [P_*(t_0)](t_1 - t_0) + 1 \right) h_{k,l}, \qquad (4.4.21)
$$

where $h_{k,l}$ represents the number of pulses at a (x_k, y_l) pixel location and the double summation gives the contribution to monomer conversion at pixel (x_i, y_j) from all pixels (x_k, y_l) scanned by the beam on the monomer surface. In matrix form, Eq. (4.4.21) can be expressed as

$$
\{ C_{k,l}(i,j) \} \{ h_{k,l} \} = \left\{ \ln \frac{[M(x_i, y_j, z)]}{[M]_0} \right\}, \qquad (4.4.22)
$$

where

$$
C_{k,l}(i,j) = -k_p \ln \left(\frac{e^{qt_0} + e^{-qt_0}}{2} \right) - \frac{k_p}{k_t} \ln \left(k_t [P_*(t_0)](t_1 - t_0) + 1 \right). \quad (4.4.23)
$$

In Eq. (4.4.22), $\{ C_{k,l}(i,j) \}$ is a $n \times n$ $(n = n_1 \times n_2)$ known matrix, $\{ h_{k,l} \}$ is a $n \times 1$ unknown column vector, and the right-hand side is a $n \times 1$ known column vector. Equation (4.4.22) is a set of n linear equations with n unknowns which can be solved using standard techniques. The solution when the right hand side is set equal to a constant, based on the specified $\frac{[M]}{[M]_0}$, gives the number of pulses $h_{k,l}$ at a pixel or the scan pattern on the surface. The solution need not be an integer. In this case, one can round to the nearest integer value. This is not expected to cause much of a problem because one will be operating at the flat portion of the curve relating percent monomer conversion to polymerization time.

The model can be readily modified to consider simultaneously both unimolecular termination caused by occlusion and bimolecular termination. For this case, Eq. (4.4.16) can be modified to give the polymerization rate

$$
-\frac{d[M]}{dt} = \beta k_p [P'_*(t)][M] + (1 - \beta) k_p [P_*(t)][M], \qquad (4.4.24)
$$

where $[P_*(t)]$ is obtained form Eqs.(4.4.14) and (4.4.15) for bimolecular termination and $[P'_*(t)]$ for unimolecular termination is obtained from the solution to the equations

$$
\frac{d[P'_*(t)]}{dt} = \phi I_a - k_t [P'_*(t)] \; ; \; 0 \le t \le t_0, \qquad (4.4.25)
$$

and

$$\frac{d[P'_*(t)]}{dt} = -k_t[P'_*(t)] \; ; \; t_0 \le t \le t_1. \tag{4.4.26}$$

Equation (4.4.24) can be solved to give $\ln\left[\frac{[M(x_i,y_j,z)]}{[M]_0}\right]$ and the $\{C_{k,l}(i,j)\}$ matrix from which one may obtain the number of pulses at a pixel.

This model combined with experiments is useful for determining the values of the operating parameters (such as beam intensity, beam radius, pixel size, percent monomer conversion, and monomer layer thickness) that give an optimum response in terms of surface smoothness and dimensional accuracy of the solidified photopolymer.

5

Laser Ablation

5.1 Introduction

Pulsed-laser-induced photoablation of polymers is a fast developing prototyping technique for the manufacturing of high aspect ratio microstructures. The process involves spontaneous ablation of surface material upon absorption of a pulse of UV laser light whose energy exceeds the material ablation threshold value. Material removal can be thermal, photochemical, or photophysical. Thermal ablation is caused by laser-induced heating, melting, and vaporization of material. Photophysical ablation is caused by non-thermal excitation like electronic excitation, while photolytic bond breaking leads to photochemical ablation. There is usually insignificant thermal damage resulting from laser ablation. This can be attributed to the short duration of the pulse, the low thermal diffusivity of polymers, and the minimization of heat build-up caused by heat energy being carried away from the surface with the ablated material. Studies on pulsed-laser-induced ablation have focused mainly on the mechanisms underlying the ablation process and on developing models to predict the etch rate or single-pulse ablated depth. Some models consider the ablation process to be photochemical, in accord with Beer's law, thermal (based on the Arrhenius equation), or a combination of thermal and photochemical. Other models are photophysical (based on activated thermal desorption of electronically excited species) or mechanical (based on stress build-up in the material). Etch rate predictive models are essential for developing models to control and optimize the photoablation system. With the help of such models, it is possible, for example, to control the scanning pattern of the excimer laser on the material surface in order to produce microstructures of pre-specified geometries.

5.2 Photothermal Models

Thermal ablation of polymers can be explained by intense heating of the polymer material, when exposed to a laser pulse, causing thermal breakdown and ejection of fragments from the surface. There are several experimental observations which have been cited in support of the thermal nature of ablation. These are the observed tail (as predicted by the Arrhenius thermal rate equation) near threshold of the curve depicting etch rate as a function of fluence (Kueper et al. 1993), the dependence of ablation threshold on the pulse repetition rate, and the fact that there has been no ablation observed without a temperature rise (Burns and Cain 1996).

A simple thermal model, assuming the existence of an energy threshold, was discussed by Furzikov (1990) in which he showed, based on Beer's law, that the ablation depth per pulse, z_d, relates to the pulse fluence by the expression

$$z_d = (\frac{B\lambda}{\alpha})^{\frac{1}{2}} \ln(\frac{F}{F_{th}}), \qquad (5.2.1)$$

where α is the linear absorption coefficient, λ is the laser wavelength, F is the laser fluence, F_{th} is the threshold fluence, and B is a constant equal to $(\sqrt{2}+1)(2\pi\sqrt{2n})^{-1}$, where n is the polymer refractive index. The threshold fluence is given by

$$F_{th} = \sqrt{2}\Delta T(\alpha)^{-\frac{1}{2}}(\pi\rho^3 C^3 \kappa t_p)^{\frac{1}{4}}, \qquad (5.2.2)$$

where ΔT is the polymer surface temperature rise, ρ is the density, C is the heat capacity, t_p is the laser pulse duration, and κ is the thermal conductivity.

Equation (5.2.1) is similar to the conventional model using Beer's law (Jellinek and Srinivasan, 1984) in which the etch depth z_d per pulse is given by

$$z_d = \frac{1}{\alpha} \ln(\frac{F}{F_{th}}). \qquad (5.2.3)$$

Equations (5.2.1) and (5.2.3) underestimate the observed etch depth for relatively large fluences. It is known that the model in Eq. (5.2.3) agrees with experimental data only in the low fluence range of less than 1 $\frac{J}{cm^2}$ (Jellinek and Srinivasan, 1984; Deutsch and Geis, 1983; Brannon et al., 1991). In an attempt to improve on the agreement between Eq. (5.2.1) and experimental data, Furzikov took into account the heat diffusion per pulse by defining an effective absorption coefficient, α_{eff}, to replace α in Eqs. (5.2.1) and (5.2.2). This is given by

$$\alpha_{eff}^{-1} = \alpha^{-1} + 2(\frac{\kappa t_p}{\rho C})^{\frac{1}{2}}, \qquad (5.2.4)$$

where $\frac{\kappa}{\rho C}$ is the thermal diffusivity. Figure (5.2.1) on etch depth in μm per pulse for poly(ethylene terephtalate), PET, shows that Eq. (5.2.1) gives better agreement with experimental data (from Lazare and Granier, 1989) when the heat diffusion is taken into consideration (α is replaced by α_{eff}). In calculating the theoretical etch depth, ΔT was $1200K$. The parameters corresponding to the polymer were $\rho = 1.41 \frac{g}{cm^3}$, $C = 1.32 \frac{J}{gK}$, $\kappa = 2.8 \times 10^{-3} \frac{W}{cmK}$, and $n = 1.66$.

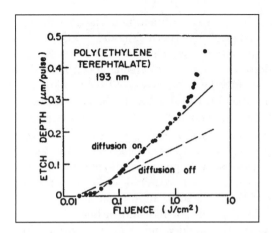

Fig. 5.2.1. Experimental (full circles) and predicted (line) etch depth per pulse as a function of laser fluence for poly(ethylene terephtalate), PET, ablation (reprinted from Furzikov 1990 with permission from the American Institute of Physics). Experimental data from Lazare and Granier 1989 with permission from Taylor & Francis http://www.tandf.co.uk/journals)

If one assumes that the concentration of the polymeric material is unaffected by material removal because of ablation, then the rate of change of polymer mass over time can be related to the Arrhenius equation and expressed as (Kueper et al., 1993)

$$\frac{dm}{dt} = A \exp(-\frac{E_a}{kT}), \tag{5.2.5}$$

where A is a constant, E_a is the activation energy, T is the surface temperature, and k is the Boltzmann constant. To obtain the change in mass per pulse duration Δt, one may assume that the temperature rise is constant and proportional to the incident fluence, ϕ. As such,

$$\Delta m = A \exp(-\frac{B}{\phi})\Delta t, \tag{5.2.6}$$

where B is a constant encompassing E_a, k, and the proportionality constant relating ϕ to T.

From Eq. (5.2.6), it is seen that the ablation per pulse increases exponentially with an increase in fluence. Evidence for this sort of exponential relation between rate and fluence was demonstrated experimentally with polyimide film ($Kapton^{TM}$) using a sensitive quartz crystal microbalance (QCM) to measure the ablation per pulse for low fluences. The QCM measures changes in the thickness of the film by recording changes in the crystal oscillation period. Figure (5.2.2) shows a fit of Eq. (5.2.6) to experimental data on the ablation rate in $\overset{0}{A}$ per pulse as a function of incident fluence in $\frac{mJ}{cm^2}$ for three excimer laser wavelengths. The thermal model in Eq. (5.2.6) fits the data very well for 248 nm and 308 nm wavelengths in the low fluence range. The Arrhenius model does not show good agreement with the data for the 193 nm wavelength, particularly near the threshold. Data plots for 248 and 308 nm do not exhibit an ablation threshold as is the case for the 193 nm plot. The 248 nm plot exhibits an exponential increase initially (between 15-35 $\frac{mJ}{cm^2}$), a linear increase between 35 and 55 $\frac{mJ}{cm^2}$, and a slight decrease in the slope of the ablation rate at higher fluence. The same is true of the 308 nm plot, except for the fact that the decrease in slope at high fluence has not been attained. This indicates that higher fluence is required at longer wavelength for the characteristic shape of the Arrhenius curve to be fully expressed. The observed ablation rate in the 193 nm plot indicates a linear relationship between rate and fluence above the threshold value. A linear relationship is consistent with a photochemical ablation mechanism. These results and other similar experimental results by the authors, indicate that perhaps the ablation mechanisms is fluence-dependent.

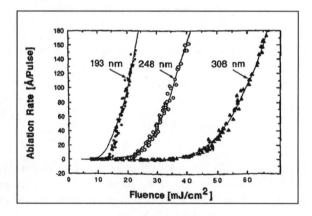

Fig. 5.2.2. Observed and predicted (lines) ablated depth per pulse for different wavelengths in the case of Kapton^{TM} (reprinted from Kueper et al. 1993 with permission from Springer-Verlag)

The temperature as a function of time and depth $(z), T(z,t)$, was calculated based on the one-dimensional heat conduction equation in the z-direction. The initial condition for solving the heat equation was $T(z,0) = T_0$, where T_0 is the ambient temperature, and the boundary conditions were $\frac{dt}{dz}|_{z=0}= 0$ and $T(\infty, t) = T_0$. For each wavelength, the maximum surface temperature, $T_{\max}(0, t)$, was calculated for the fluence at which ablation could be first detected. These fluences were 14, 19, 37, and 84 $\frac{mJ}{cm^2}$ for wavelengths 193, 248, 308, and 351, respectively. Also, the maximum temperatures were 819, 878, 833, and 877 C^0 for wavelengths 193, 248, 308, and 351 nm, respectively. It was argued by the authors that the similarity of the maximum temperatures over the different wavelengths supports the hypothesis of a photothermal ablation mechanism. Also, if the ablation mechanism at 193 nm wavelength was indeed photochemical, it seemed to commence when the surface temperature was at a maximum.

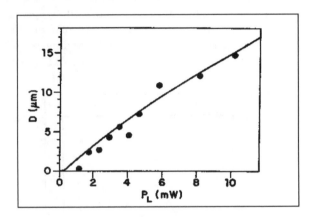

Fig. 5.2.3. Predicted (solid line) and observed etch depth as a function of laser power for polyimide. Etch time was 1 s, laser wavelength was 257 nm, and ambient was oxygen at one atmosphere (reprinted from Treyz et al. 1989 with permission from the American Institute of Physics)

Experimental observations on etch depth for high aspect ratio features in a polyimide film, using an Ar^+ cw laser at relatively low power of $2-12\ mW$, were consistent with a thermal mechanism based on the Arrhenius relation (Treyz et al. 1989)

$$\frac{dz}{dt} = A\exp(-\frac{E_a}{kT(z)}), \qquad (5.2.7)$$

where the temperature at the center of the etched surface at depth z in the case of a Gaussian beam is given by

$$T(z) = \frac{\eta(1 - R)P}{2\sqrt{\pi}\kappa\omega(z)} + T_0, \qquad (5.2.8)$$

where $\omega(z) = \omega_0 + az$. Here, η is the throughput of the optical system, P is the laser power at the surface, R is reflectivity, T_0 is the ambient temperature,

κ is thermal conductivity, a is a constant, and $\omega(z)$ is the $\frac{1}{e}$ laser beam radius at depth z. Combining Eqs. (5.2.7) and (5.2.8) and integrating over time gives the function for etch depth

$$z(t) = \frac{\gamma P}{E_a} \ln \left[\left(\frac{E_a}{\gamma P} \exp(-\frac{E_a \omega_0}{\gamma Pa}) At \right) + 1 \right], \qquad (5.2.9)$$

where $\gamma = \frac{k\eta(1-R)}{2\sqrt{\pi}\kappa a}$. Equation (5.2.9) was derived under the assumption that the maximum temperature rise at the center of the irradiated spot is much higher than the ambient temperature. Because of the approximate nature of Eq. (5.2.9), numerical integration was used to fit the model to data on etch depth in a polyimide film (Dupont 2555). Model parameters used were $E_a = 0.84eV$, $\eta = 0.29$, $A = \frac{1.0\ cm}{s}$, $R = 0.12$, $\kappa = 6.2 \times 10^{-4} \frac{W}{cmC^0}$, $\omega_0 = 1.8\mu m$, $a = 3.0$, etch time $= 1s$, and Ar^+ laser wavelength $\lambda = 257nm$. Figures (5.2.3) and (5.2.4) showed good agreement between model and measured etch depth as a function of laser power and the logarithm of etch time, respectively.

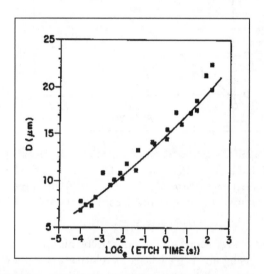

Fig. 5.2.4. Predicted (solid line) and observed etch depth as a function of log etch time in seconds. Laser power was 10 mW, laser wavelength was 257 nm, and ambient was oxygen at one atmosphere (reprinted from Treyz et al. 1989 with permission from the American Institute of Physics)

Thermal models based on the Arrhenius relation or Beer's law as in Eqs. (5.2.1) and (5.2.3) seem to fit etch depth data in the low fluence range. However, at higher fluence, there are substantial deviations in fit between these models and experimental etch depth measurements.. Lack of agreement between model and data can be largely eliminated by taking into account the thermal diffusion per pulse from the surface to the material bulk. Inclusion of thermal diffusion into the Arrhenius model was discussed by Cain et al. (1992 a,b). The etched depth z, as in Trez et al. (1989), is given by Eq. (5.2.7). In this case, however, the temperature $T(z)$ is calculated based on Beer's law of absorption to be

$$T(z) = \frac{\alpha F}{C_p} \exp(-\alpha z), \tag{5.2.10}$$

where C_p is the specific heat capacity per unit volume $(\frac{J}{K cm^3})$, F is the fluence of the incident pulse, and α is the linear absorption coefficient. The effect of thermal diffusion was investigated by treating $T(z)$ in Eq. (5.2.10) as an initial condition and calculating the new temperature distribution based on the diffusion equation

$$D \frac{\partial^2 T}{\partial z^2} = \frac{\partial T}{\partial t}, \tag{5.2.11}$$

where D in $cm^2 s^{-1}$ is the thermal diffusivity. To simulate the etch process, the initial surface temperature from Eq. (5.2.10) is used to solve for $T(z)$ from Eq. (5.2.11), subject to appropriate boundary conditions. This $T(z)$ temperature in combination with Eq. (5.2.7) gives the new depth z. Based on the new depth, Eq. (5.2.10) gives the initial condition for the new surface from which the new temperature $T(z)$ may be obtained from the solution to Eq. (5.2.11). This iteration is continued until a given time t or until the change in surface position goes to zero. Although thermal diffusion seemed to account for the observed deviations from etching data, by giving an effective absorption coefficient smaller than α in Eqs. (5.2.1) and (5.2.3), it gave rise to unrealistically high temperatures (Cain et al. 1992a) suggesting that the etching mechanism does not conform to the mode of operation described above. An improvement on the model was obtained (Cain et al. 1992b) by assuming a first-order thermal degradation of the polymer based on the formula

$$\frac{dn}{dt} = -k(T)n, \tag{5.2.12}$$

where n is the number of single bonds in the polymer. The rate constant was expressed as a function of temperature according to the equation

$$k(T) = \frac{kT}{h} \exp(-\frac{E_a}{kT})., \tag{5.2.13}$$

where h is Planck's constant. Here, it is assumed that ablation occurs when the number of bonds n drops below a predetermined proportion, n_p, of the original number. In this model, the temperature is calculated from the absorbed photon density by dividing by the specific heat.

Photon absorption was modeled based on chromophore transition between three states, two excited states, and the ground state. Let n_g, n_1, and n_2, denote the densities (cm^{-3}) of chromophores in the ground state, first excited state, and second excited state, respectively. Photon absorption, assuming equal energy space between first and second state, is governed by the set of equations

$$\frac{dn_g}{dt} = \sigma_1(n_1 - n_g)p + \sigma_{12}(n_2 - n_g)p^2 + \frac{n_1}{\tau_1} + \frac{n_2}{\tau_2},$$

$$\frac{dn_1}{dt} = \sigma_1(n_g - n_1)p + \sigma_2(n_2 - n_1)p - \frac{n_1}{\tau_1}, \qquad (5.2.14)$$

$$\frac{dn_2}{dt} = \sigma_2(n_1 - n_2)p + \sigma_{12}(n_g - n_2)p^2 - \frac{n_2}{\tau_2},$$

where $n_g + n_1 + n_2 = N = $ constant, p is the photon flux in $cm^{-2}s^{-1}$, τ_1 and τ_2 are the lifetime of the first and second states, σ_1 and σ_2 are the absorption cross-sections in cm^2 for the transition from ground to the first excited state and from the first to the second excited state, and σ_{12} is the absorption cross-section (cm^4) for the transition from ground to the second excited state. The linear absorption coefficient can be expressed as $\alpha = \sigma_1 N$. Beer's law applies if all N chromophores are in the ground state, which can occur under low pulse fluence or long pulse duration. Solution to the set of equations in (5.2.14) was accomplished numerically by dividing the polymer into a collection of 1,000 thin slabs of material and dividing the laser pulse into temporal segments. The density of absorbed photons in a given slab was calculated as

$$d_{abs} = \int_0^{\delta t} \left[\sigma_1(n_g - n_1)p + \sigma_2(n_1 - n_2)p + 2\sigma_{12}(n_g - n_2)p^2\right] dt, \quad (5.2.15)$$

where $\delta t = \frac{\text{pulse length}}{\text{number of temporal segments}}$. The equivalent heat in the ith slab was calculated as $T_i = \frac{h\nu d_{abs}}{C_p}$ (details are in Cain et al. 1992 b). The heat generated from photon absorption was used as initial condition, and the temperature distribution generated from the diffusion equation in (5.2.11) was applied to determine thermal degradation at various positions of the polymer from Eqs. (5.2.12) and (5.2.13). This iteration was continued until the number of bonds reached below n_p, where ablation was assumed to have occurred. The polymer degradation calculations were continued until the final temperatures became low and the etch depth remained unchanged. In this model, it can be shown that the initial temperature as calculated from the chromophore photon absorption is lower and more reasonable in value than that based on Eq. (5.2.10). This model can explain the deviations of the models based on Beer's law from observed etch depth at high fluence. Figure (5.2.5) gives the temperature profile for polyimide as a function of depth for different fluences. It is clear that, for a high fluence of 5.0 J cm^{-2}, the temperature decay deviates considerably from an exponential (Beer's law). This implies that the present model can simulate the observed etch depth at high fluence, which is known to be larger than that predicted from models based on Beer's law. In fact, using the same model (Cain 1993) with the added assumption of chromophore modification (where two types of chromophores, normal and incubated, are considered) according to the first-order kinetics in Eqs. (5.2.12) and (5.2.13), good agreement between model and experimental observations for etch depth were observed at high fluence.

To study the effect of pulse repetition rate on the relation between etch rate and fluence (for low fluence where Beer's law applies), the heat diffusion equation (5.2.11) was solved analytically with Eq. (5.2.10) serving as initial condition (Burns and Cain 1996). The solution for the temperature rise is given by

$$T(z,t) = \frac{\alpha F}{C} \exp(\alpha^2 Dt)[\exp(-\alpha z) \, erfc(\alpha\sqrt{Dt} - \frac{z}{2\sqrt{Dt}})$$
$$+ \exp(\alpha z) \, erfc(\alpha\sqrt{Dt} + \frac{z}{2\sqrt{Dt}})], \qquad (5.2.16)$$

where $erfc$ is the complimentary error function. The surface temperature rise can be expressed as

$$T_s = \frac{\alpha F^*}{C} \exp(\alpha^2 D\tau) \, erfc(\alpha\sqrt{D\tau}). \qquad (5.2.17)$$

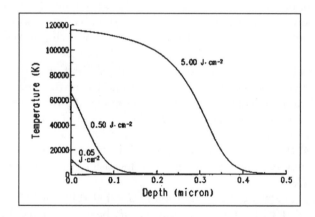

Fig. 5.2.5. Predicted temperature profiles as a function of depth for different laser fluences. Pulse length was 1.5 ns, and laser wavelength was 193 nm (reprinted from Cains et al. 1992b with permission from the American Institute of Physics)

Since the time between pulses is large relative to the pulse duration (τ), t is replaced by τ in Eq. (5.2.17). Considering pulses in sequence, one may express the surface temperature T_s^{n+1} after absorption of the nth pulse in terms of that before absorption. This gives

$$T_s^{n+1} = T_s^n + \frac{\alpha F^*}{C} \exp(\alpha^2 D\tau) \, erfc(\alpha\sqrt{D\tau}). \qquad (5.2.18)$$

It is seen that after N pulses with a repetition rate of v hertz, the surface temperature is given by

$$T_s^{N+1} = T_s^0 + \sum_{n=1}^{N} \frac{\alpha F^*}{C} \exp(\frac{\alpha^2 Dn}{v}) \, erfc(\frac{\alpha\sqrt{Dn}}{\sqrt{v}}), \qquad (5.2.19)$$

where $T_s^0 = T_s^1$ is the ambient temperature when $n = 0$. Here, F^* is taken to be the threshold fluence since material with thermal energy above the threshold is removed and does not contribute to the re-distribution of energy. In order to determine the etch rate, the simulation method (Cain et al. 1992b) involving Eq. (5.2.11)-(5.2.13), as explained above, was used. In this case, however, the initial condition was obtained by calculating, for each pulse, a surface temperature T_s from Eq. (5.2.19) and adding it to $T(z)$ from Eq. (5.2.10). Results indicated that the curve depicting etch rate versus fluence shifted progressively to lower fluence as the repetition rate increased. This shift did not significantly change the shape of the curve. The magnitude of this parallel shift is influenced by thermal diffusivity, the number of pulses, and the absorption coefficient. For polyimide, a high absorbing polymer, the shift was not noticeable until a repetition rate of 100 Hz and was of small magnitude (about $.01\frac{J}{cm^2}$ in the fluence threshold). On the other hand, the effect of repetition rate on the etch rate in a PMMA-based polymer (Vacrel^{TM} 8230), which has a low absorption coefficient relative to polyimide, was pronounced. There was a decrease in fluence threshold (shift in the etch rate versus fluence curve to lower fluence) from 1.012 $\frac{J}{cm^2}$ at 10 Hz to 0.95 $\frac{J}{cm^2}$ at 100 Hz and to 0.667 $\frac{J}{cm^2}$ at 300 Hz. Reduction in the fluence ablation threshold can be explained as being caused by the accumulation of heat from previous pulses and can become significant at repetition rates above 100 Hz for low absorbing polymers. Further results indicated that the etch depth in polyimide was found to increase linearly with the number of pulses at low repetition rates of 10 Hz or less. At 100 Hz, the increase in etch depth became slightly nonlinear. This nonlinearity may become even more pronounced for low absorbing polymers such as PMMA.

A thermal model taking into account thermal diffusion where the heat source is coupled with photon absorption has been considered by D'Couto and Babu (1994). It was assumed as in Kueper et al. (1993) that material ablation commences once a threshold temperature is reached. Heat transfer in the polymer was modeled using the one-dimensional heat transfer equation in rectangular coordinates.

$$\frac{\partial^2 T}{\partial x^2} = \frac{\partial T}{\kappa \partial t} - \frac{A(x,t)}{k}, \qquad (5.2.20)$$

where x is the depth perpendicular to the ablation front, $A(x,t)$ is the heat source, κ is the thermal diffusivity, and k is the thermal conductivity. The boundary conditions at different depths are $\frac{\partial T}{\partial x} = 0$, and $T \rightarrow T_{ambient}$ as

$x \rightarrow \infty$. The initial condition is $T = T_{ambient}$ at $t = 0$. The heat source is given by

$$A(x,t) = \rho_1(x,t)hv, \tag{5.2.21}$$

where h is the Planck's constant, v is the frequency of the incident radiation, and ρ_1 is the chromophore concentration of the first excited state. The single-photon absorption dynamics model proposed by Pettit and Sauerbrey (1993) was used to calculate ρ_1. Considering a two-state (ground state and first excited state) model, the concentration of the first excited state chromophore is expressed as

$$\frac{\partial \rho_1(x,t)}{\partial t} = -\frac{\partial \rho_0(x,t)}{\partial t} = \sigma_1[\rho_0(x,t) - \rho_1(x,t)]S(x,t) - \frac{\rho_1(x,t)}{\tau}, \tag{5.2.22}$$

where σ_1 is the single-photon absorption cross-section, $S(x,t)$ is the photon flux, ρ_0 is the chromophore concentration in the ground state, and τ is the decay time of the excited state. At $t = 0$, ρ_1 can be assumed to be zero. The attenuation of the incident photon flux with depth is given by

$$\frac{dS}{dx} = -\frac{\rho_0}{2}(1 - e^{-2\sigma_1 S(x,t)}). \tag{5.2.23}$$

Equations (5.2.20), (5.2.22), and (5.2.23) were solved numerically using finite difference techniques. The laser pulse was assumed to be rectangular in shape and was discretized in time. The number of time segments were $60 - 100$, depending on the fluence. The time interval for a segment was such that, starting from the ambient temperature, the heat generated from one pulse segment was not sufficient to raise the polymer temperature above the threshold temperature. The chromophore concentration $\rho_1(x,t)$ for calculating the heat source from Eq. (5.2.21) was obtained from the finite difference solution to Eqs. (5.2.22) and (5.2.23). Considering the pulse segments sequentially and starting with the first segment, polymer regions that attain the threshold temperature were assumed to be degraded. The depth of degraded material, x_d, was allowed to increase on further exposure to subsequent segments within the same pulse. The photon flux reaching the undegraded polymer at depth x_d was expressed as

$$S(x_d, t) = S(0, t)[1 - \exp(-\alpha_d x_d)], \tag{5.2.24}$$

where α_d is an effective absorption coefficient, assumed to be a constant. Before the start of the next pulse, all degraded material was assumed to have left the surface. Also, residual heat in the undegraded material was assumed not to have any effect on subsequent ablation during the next pulse. In fitting the above model to experimental data on etch rate, the effective absorption coefficient α_d of the degraded material was the only model parameter to consider. This parameter was set equal to the function AF^B, where F is the fluence and A and B are constants determined by fitting the function to fluence-dependent absorption data. There were very good agreements between model predictions and experimental data on etch rate as a function of fluence in the case of polyimide (Kapton-H), PMMA, PEEK, PES, and PET exposed to 248 nm and 308 nm pulse wavelengths and a range of pulse durations ($21ns - 231ns$). Figure (5.2.6) is one such fit for polyimide. It shows excellent agreement with the observed etch rate in $\frac{\mu m}{pulse}$ for a wide fluence range.

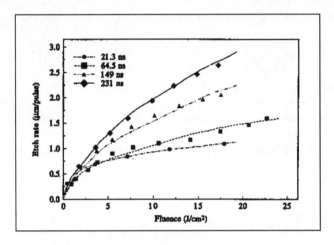

Fig. 5.2.6. Prediction of etch depth per pulse for polyimide at 248 nm wavelength and for different pulse durations. Experimental data from Schmidt et al 1998 (reprinted from D'Couto and Babu 1994 with permission from the American Institute of Physics)

A thermal UV laser ablation model, which considers absorption of incident beam by the ablated particles and a moving interface, has been proposed by Tokarev et al. (1995). The model takes into account heat flow, in the z-direction from the surface into the material volume, according to the heat conduction equation with a moving interface. As such, the temperature profile inside the material is described by the unsteady state convective-diffusion equation

$$\frac{\partial T(z,t)}{\partial t} = \kappa \frac{\partial^2 T(z,t)}{\partial z^2} + v(t)\frac{\partial T(z,t)}{\partial z}$$

$$+ \frac{(1-R)I(t)\alpha}{C} \exp(-\alpha z), \qquad (5.2.25)$$

with

$$T(z,0) = T(\infty,t) = T_i \; ; T(0,t) = T_s(t),$$

$$and \quad \frac{\partial T(0,t)}{\partial z} = \frac{L}{\kappa C}v(t), \qquad (5.2.26)$$

where R is surface reflectivity, $v(t)$ is the velocity at which the interface moves into the depth of the material as a result of laser-induced ablation, C is the heat capacity, L is the latent heat of vaporization per unit volume, α is the absorption coefficient, κ is thermal diffusivity, T_i is the initial temperature, and T_s is the surface temperature. At steady state, with $I(t) = I_0 = \text{constant}$, $v(t) = v_0 = \text{constant}$, and $T_s = T_0 = \text{constant}$, Eq. (5.2.25) becomes

$$\kappa \frac{\partial^2 T(z)}{\partial z^2} = -v_0 \frac{\partial T(z)}{\partial z} - \frac{(1-R)I_0\alpha}{C}\exp(-\alpha z). \tag{5.2.27}$$

The solution to Eq. (5.2.27) is given by

$$T(z) = A\exp(-\alpha z) + B\exp(-\frac{v_0 z}{\kappa}) + T_i, \tag{5.2.28}$$

where

$$A = \frac{(1-R)I_0}{(v_0 - \alpha\kappa)C}, \quad B = -\frac{(1-R)I_0\alpha\kappa}{(v_0 - \alpha\kappa)Cv_0} - \frac{L}{C}. \tag{5.2.29}$$

The boundary conditions in Eq. (5.2.26) satisfy the solution in Eq.(5.2.28) when

$$v_0 = \frac{(1-R)I_0}{C(T_0 - T_i) + L}. \tag{5.2.30}$$

When I is not constant, but a function of t, the interface velocity $v(t)$ assuming a quasistationary process was expressed, based on Eq. (5.2.30), as

$$v(t) = \frac{(1-R)I(t)}{C(T_0 - T_i) + L}; \; t > t_{th}, \tag{5.2.31}$$

where t_{th} is a time threshold at which the surface temperature reaches the steady state value T_0. When absorption of the laser beam by the ablated particles (plasma) is taken into effect, the beam intensity $I(t)$ transmitted into the plasma can be expressed as

$$I(t) = I_0(t)\exp[-\int_{-H(t)}^{z(t)} \beta(z',t)dz']. \tag{5.2.32}$$

Here, $\beta(z,t)$ is the absorption coefficient of the plume or plasma. The limits of integration change with time as a result of plume expansion and the moving interface, $z(t)$, which can be expressed as $z(t) = \int_{t_{th}}^{t} v(t')dt'$. From Eqs. (5.2.31) and (5.2.32), one has that

$$v(t) = \frac{dz}{dt} = \frac{(1-R)}{C(T_0 - T_i) + L}I_0(t)\exp[-\int_{-H(t)}^{z(t)} \beta(z',t)dz']. \tag{5.2.33}$$

It is seen also that

$$\alpha z(t) = \sigma_0 N(t); \; and \; \int_{-H(t)}^{z(t)} \beta(z',t)dz' = \sigma_p(t)N(t), \tag{5.2.34}$$

where σ_0 and σ_p are the absorption cross-sections for the polymer material and the plasma, respectively, and $N(t)$ is the total number of particles in a material volume of depth $z(t)$ and area 1 cm^2. Equation (5.2.34) gives the relation

$$\frac{\int_{-H(t)}^{z(t)} \beta(z',t)dz'}{\alpha z(t)} = \frac{\sigma_p(t)}{\sigma_0} = \mu(t). \tag{5.2.35}$$

In what follows, $\sigma_p(t)$ and hence $\mu(t)$ were considered constant over time. This was considered valid for UV laser ablation when absorption in the plasma is caused mainly by single-photon ionization of plasma particles. With μ constant, Eq. (5.2.33) becomes

$$\frac{dz}{dt} = \frac{(1-R)}{C(T_0 - T_i) + L} I_0(t) \exp[-\mu\alpha z]. \tag{5.2.36}$$

Equation (5.2.36) can be integrated to give

$$\int_0^{z(t)} \exp(\mu\alpha z')dz' = \frac{(1-R)}{C(T_0 - T_i) + L} \int_{t_{th}}^t I_0(t')dt', \tag{5.2.37}$$

which yields the expression for etch depth

$$z(t) = \frac{1}{\mu\alpha} \ln\left(1 + \frac{(1-R)\mu\alpha}{C(T_0 - T_i) + L}[E(t) - E_{th}]\right). \tag{5.2.38}$$

Here, $E(t) = \int_0^t I_0(t')dt'$ and $E_{th} = \int_0^{t_{th}} I_0(t')dt'$ are the surface incident beam intensities at times t and t_{th}. The etch rate or thickness of material ablated per pulse is seen from Eq.(5.2.38) to be

$$z = \frac{1}{\mu\alpha} \ln\left(1 + \frac{(1-R)\mu\alpha}{C(T_0 - T_i) + L}[E - E_{th}]\right), \tag{5.2.39}$$

where E is the intensity per pulse. Further analysis of Eq. (5.2.39), when R and α were temperature-dependent, gave

$$z = \frac{1}{\mu\alpha} \ln(1 + \mu\delta\frac{E - E_{th}}{E_{th}}), \tag{5.2.40}$$

where

$$\delta = \left(\int_{T_i}^{T_0} \frac{C}{(1 - R(T))\alpha(T)}dT + \frac{L}{(1 - R(T_0))\alpha(T_0)}\right) \times \tag{5.2.41}$$
$$\frac{(1 - R(T_0))\alpha(T_0)}{C(T_0 - T_i) + L}$$

Expression (5.2.40) was examined under different scenarios for $\mu\delta$ (less than, equal, or greater than one) in order to illustrate the relations between the etch rate z and E. It was shown that relations generated were similar to what has been reported in the literature from ablation experiments, thus supporting the validity of the model. For instance, when $\mu\delta\frac{E-E_{th}}{E_{th}}$ is much smaller than 1, Eq. (5.2.40) reduces to the linear relation, $z = \frac{\delta(E-E_{th})}{\alpha E_{th}}$. Also, when $\mu\delta = 1$, z becomes linearly dependent on $\ln E$, $z = \frac{\delta}{\alpha}\ln\frac{E}{E_{th}}$. Both linear and log-linear relations between experimentally observed etch rate and fluence have been reported in the literature.

Photothermal models consider ablation mainly as a surface process. Few models have considered bulk chemical reactions which may produce volatile by-products within the polymer material. These by-products can give rise to internal stresses that can influence the ablation process. One such model was considered by Bityurin et al. (1998) in which laser ablation was related to the simple thermally activated reaction

$$A \rightarrow B. \tag{5.2.42}$$

In this formulation, A stands for the material bulk and B for the by-products resulting from broken bonds. The governing equations describing the ablation process are given by

$$\frac{\partial N_A}{\partial t} = V\frac{\partial N_A}{\partial z} - A_1 N_A \exp(\frac{-E_a}{T}), \tag{5.2.43}$$

$$\frac{\partial T}{\partial t} = V\frac{\partial T}{\partial z} + D\frac{\partial^2 T}{\partial z^2} + \frac{\alpha I}{c_p\rho}, \tag{5.2.44}$$

and

$$\frac{\partial I}{\partial z} = -\alpha I. \tag{5.2.45}$$

Here, α is the absorption coefficient of the material, I is the light intensity, c_p is specific heat, A_1 is the reaction rate, ρ is density, D is heat diffusivity, N_A is concentration of species A ($N_A + N_B = N_0$), T is absolute temperature, and V is velocity of the ablation front. Replacing αI in Eq. (5.2.44) by $-\frac{\partial I}{\partial z}$ and integrating over z gives, under steady state,

$$D\frac{\partial T}{\partial z} = \frac{I}{c_p\rho} - VT. \tag{5.2.46}$$

Also, under steady state, Eq. (5.2.43) becomes

$$V\frac{\partial N_A}{\partial z} = A_1 N_A \exp(\frac{-E_a}{T}). \tag{5.2.47}$$

Equations (5.2.46) and (5.2.47) were solved for the velocity V using the following conditions

$$\frac{\partial T}{\partial z}\big|_{z=0} = 0; \ T(z \rightarrow \infty) = T_\infty = 0; \ N_A(z \rightarrow \infty) = N_0;$$

$$N_A(0) = N_0 - N_B(0, t) = N_0 - N^*. \tag{5.2.48}$$

Solution to Eq. (5.2.47), with the conditions $N_A(\infty)$, $N_A(0)$ from Eq. (5.2.48), is given by

$$V = \frac{A_1}{\ln \frac{N_0}{N_0 - N^*}} \int_0^\infty \exp\left(\frac{-E_A}{T(z)}\right) dz. \tag{5.2.49}$$

From the condition $\frac{\partial T}{\partial z}|_{z=0} = 0$, it is readily seen from Eq. (5.2.46) that

$$T(0) = \frac{I_0}{c_p \rho V} \; ; \; \frac{\partial^2 T}{\partial z^2}|_{z=0} = -\frac{I_0 \alpha}{D c_p \rho}. \tag{5.2.50}$$

Approximating the integral by using the Taylor series expansion taking into account Eq. (5.2.50) and the condition $\frac{\partial T}{\partial z}|_{z=0} = 0$, Eq. (5.2.49) gives

$$V = \left(\frac{A_1}{\ln \frac{N_0}{N_0 - N^*}}\right)^{\frac{2}{3}} \left(\frac{T(0) D \pi}{2 E_a \alpha}\right)^{\frac{1}{3}} \exp\left(-\frac{2 E_a}{3 T(0)}\right). \tag{5.2.51}$$

The authors related this velocity expression to the velocity obtained from a surface evaporation model (Luk'yanchuk et al., 1997) namely,

$$V = V_0 \exp(-\frac{E_{a,s}}{T(0)}), \tag{5.2.52}$$

which indicates that the present model is similar to the surface evaporation model with $E_{a,s} = \frac{2 E_a}{3}$.

Further analysis considering the reaction

$$A \to B \to C, \tag{5.2.53}$$

led to the conclusion that, if the activation energy E_A, representing bond breaking, is larger than E_B (activation energy for further reaction or modification), no stationary solution exists at relatively low light intensity. On the other hand, one stable stationary solution exists for light intensities higher than a certain threshold. This finding was used as a possible explanation of experimental results on ablation from Srinivasan et al. (1995) and Lazare et al. (1996).

Further analyses of essentially the same bulk model of Bityurin et al. (1998) were considered by Arnold and Bityurin (1999). This analysis revealed certain ablation characteristics that distinguished the bulk model from surface models. Salient among them were the following: (1) sharp onset of ablation unlike the so-called Arrhenius tail observed for surface models; (2) this sharp onset is accompanied by a very high velocity resembling that of an explosion; (3) fast transition to the quasi-stationary ablation regime occurs; (4) Etch depth, d, near threshold is proportional to the square root of the laser fluence $(d \propto (\phi - \phi_{th})^{\frac{1}{2}})$.

A photothermal surface evaporation model where heat source is attributable to photon absorption by chromophores, within the framework of two and three electron levels, was developed by Bityurin and Malyshev (1998) to study the effect of delay time between two consecutive ultra short laser pulses on etch depth. Experimental conditions were considered in which the delay time between pulses is shorter than the heat diffusion time. For the three electron levels, the governing equations for the densities of chromophores in the second (n_2) and third (n_3) excited states were expressed as

$$\frac{\partial n_2}{\partial t} = V\frac{\partial n_2}{\partial z} + \frac{\sigma_{12}}{h\upsilon}I(n_1 - n_2) - \frac{n_2}{t_{21}} - \frac{\sigma_{23}}{h\upsilon}I(n_2 - n_3) + \frac{n_3}{t_{32}}, \qquad (5.2.54)$$

and

$$\frac{\partial n_3}{\partial t} = V\frac{\partial n_3}{\partial z} + \frac{\sigma_{23}}{h\upsilon}I(n_2 - n_3) - \frac{n_3}{t_{32}}, \qquad (5.2.55)$$

where $n_1 + n_2 + n_3 = n_0 =$ chromophore concentration, σ_{ij} is the transition cross-section from i to j, and t_{ij} is the transition time from state i to state j. The intensity distribution as a function of depth at time t $(I(z,t))$ was given by

$$-\frac{\partial I}{\partial z} = \sigma_{12}I(n_1 - n_2) + \sigma_{23}I(n_2 - n_3). \qquad (5.2.56)$$

The heat diffusion equation considered was

$$\frac{\partial T}{\partial z} = V\frac{\partial T}{\partial z} + \frac{1}{\rho c_p(T)}\frac{\partial}{\partial z}\left[\kappa(T)\frac{\partial T}{\partial z}\right] + \frac{h\upsilon}{\rho c_p(T)}\left(\frac{n_2}{t_{21}} + \frac{n_3}{t_{32}}\right). \qquad (5.2.57)$$

For surface ablation, the velocity of the moving interface (ablation velocity) was expressed as

$$V = V_0 \exp\left(-\frac{E_a}{k_B T_s}\right), \qquad (5.2.58)$$

where V_0 is a constant taken close to sound velocity ($\frac{10^6 cm}{s}$ in the case of polyimide), k_B is the Boltzmann constant, E_a is the activation energy, and T_s is the surface temperature obtained from Eq. (5.2.57) at $z = 0$. The boundary conditions for Eq. (5.2.57) are

$$\kappa(T)\frac{\partial T}{\partial z}\mid_{z=0} = V\rho(\Delta H); \quad T(z \to \infty) = T_\infty, \qquad (5.2.59)$$

where ΔH is the heat enthalpy for sublimation. The boundary condition for Eq. (5.2.56), taking into consideration the filtering of the light by the plume, is given by

$$I(z = 0) = I_0(t)\exp(-\alpha_g \int_0^t V(t')dt'), \qquad (5.2.60)$$

where α_g is the absorption coefficient of the plume and the integral is the ablated depth at time t. The initial conditions at $t = 0$ are $n_1 = n_0$, $n_2 = n_3 = 0$.

Solving Eqs. (5.2.54)-(5.2.57) based on the boundary and initial conditions, one obtains the surface temperature and hence the ablation velocity and ablation depth from Eq. (5.2.58). Based on the parameters used for model evaluation, it was shown that, for ultra short pulse time in fs, the ablation depth could decrease or increase with an increase in delay time (in ps) between two pulses. Here, the maximum delay time considered was shorter than the heat diffusion time. An increase in ablated depth with an increase in delay time was attributed to a process in which ablation is dependent on the surface temperature. On the other hand, a decrease in ablated depth with an increase in delay time was attributed to a process where ablation depends on the energy deposition length. Fitting the model equations to experimental data on polyimide, shows a quadratic (dampened increase) curve of ablation depth vs delay time, which indicates that the ablation is surface temperature-dependent. These results are similar to those obtained by Luk'yanchuk et al. (1997). Under a similar thermal model and ablation velocity, they showed that the ablated depth can be an increasing or decreasing function of delay time, depending on the ratio $s = \frac{\sigma_{23}}{\sigma_{12}}$, where σ_{12} is the cross-section for the transition from ground state to first excited state and σ_{23} from the first to the second excited state. For $s = 0$, the ablated depth decreased with an increase in delay time. On the other hand, for $s = 3$, the ablated depth increased with an increase in delay time. In the case of the photophysical model by Luk'yanchuk et al. (1996, 1997), the ablation velocity was expressed as

$$V = (1 - \frac{N_s^*}{N})V_0 \exp(-\frac{T_a}{T_s}) + \frac{N_s^*}{N}V_0^* \exp(-\frac{T_a^*}{T_s}), \qquad (5.2.61)$$

where N_s^* is the concentration of excited species at the surface, and T_a^*, T_a are the activation temperatures for ground and excited states, respectively. Using this photophysical model, it was shown that the ablated depth, in contrast to the thermal model, did decrease with a decrease in delay time whether $s = 0$ or 3. However, the ablated depth was substantially larger for $s = 0$ than for $s = 3$. As for the thermal model, this photophysical model can explain the observed ablated depth as a function of delay time for polyimide.

5.3 Photochemical Models

Absorbed photon energy from the laser light, if high enough, causes excitation which can break chemical bonds in a polymer chain. As a result, fragments are generated which are small enough to be ejected from the surface. This ablation process is termed photochemical and can occur without any rise in temperature. One of the first photochemical models was that presented by Jellinek and Srinivasan (1984) in which they showed that random breakage of chain links caused by absorbed photon energy leads to the empirical relation in Eq. (5.2.3), based on Beer's law, predicting etch depth as a function of fluence.

The rate of random breakage of chain links at depth z in a cross-section $A = 1$ cm^2 when the pulse duration is τ can be expressed as

$$\frac{dn_\tau}{dt} = kI_z = k\frac{F}{\tau} \; links \; cm^{-2} \; s^{-1}, \tag{5.3.1}$$

where k is a rate constant in links per mJ, I_z is the light intensity at depth z in mJ per cm^2 s, and F is the corresponding fluence in mJ per cm^2. According to Beer's law, $I_z = I_0 \exp(-\alpha z)$ and Eq. (5.3.1) reduces to

$$\frac{dn_\tau}{dt} = kI_0 \exp(-\alpha z) = k\frac{F_0}{\tau} \exp(-\alpha z). \tag{5.3.2}$$

Integrating Eq. (5.3.2) over the time interval $(0, \tau)$ gives the number of broken chain links in a volume element for the pulse duration

$$n_\tau(z) = kI_0 \exp(-\alpha z)\tau = kF_0 \exp(-\alpha z), \tag{5.3.3}$$

where $I_0 \exp(-\alpha z)\tau$ or $kF_0 \exp(-\alpha z)$ represent the intensity or fluence per cm^2 that entered the volume element in time τ seconds. It is clear that the volume ejected, dz, is proportional to the number of broken links. As such, one has

$$n_\tau(d) = k_v dz = kI_0 \exp(-\alpha z)\tau = kF_0 \exp(-\alpha z), \tag{5.3.4}$$

where k_v is the proportionality constant (cm^{-3}). Integrating Eq. (5.3.4) over the interval $(0, z_d)$ gives

$$(e^{\alpha z_d} - 1) = \frac{\alpha k}{k_v}(F_0 - F_{th}) = \frac{F_{th}\alpha k}{k_v}\left(\frac{F_0}{F_{th}} - 1\right), \tag{5.3.5}$$

where the threshold value F_{th} is the fluence at depth z_d. Replacing F_{th} in Eq. (5.3.5) by $F_0 e^{-\alpha z_d}$ gives the relation

$$1 = \frac{F_{th}\alpha k}{k_v}. \tag{5.3.6}$$

From Eqs. (5.3.5) and (5.3.6), one obtains the etch depth as a function of fluences

$$z_d = \frac{1}{\alpha} \ln\left(\frac{F_0}{F_{th}}\right). \tag{5.3.7}$$

This expression is also valid for long-term irradiation over time, t. In this case, $F_0 = I_0 t$. Similar results were obtained when chain degradation is not through random breakage, but through depolymerization where the chain length is reduced in length one monomer at a time starting from one end or the other. Depolymerization of a chain commences after a chain end absorbs a photon to produce a chain end radical. In this case, the etch depth per pulse is given by

$$z_d = \frac{1}{\alpha'} \ln\left(2N\frac{F_0}{F'_{th}}\right), \tag{5.3.8}$$

where N is the number of chain links broken for each chain molecule.

Figure (5.3.1) presents a fit of the model in Eq. (5.3.7) to data, on etch depth as a function of $\ln(F_0)$, from experiments involving poly(methyl methacrylate) or PMMA; polycarbonate; poly(ethylene terephthalate) or PET (Mylar); and polyimide (Kapton). Here, $I_f = z_d$ from Eq. (5.3.7). The model shows excellent fit to data for low fluence (near threshold). It is interesting to note that the thresholds indicate a sharp onset of ablation and do not exhibit the Arrhenius tail characteristic of photothermal surface models (Fig. 5.2.2). However, sharp onset of ablation was shown to occur for a photothermal bulk model (Arnold and Bityurin 1999).

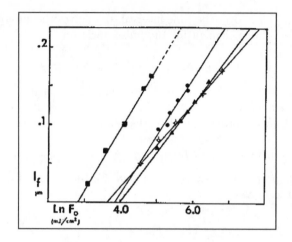

Fig. 5.3.1. Predicted and observed etch depth for PMMA, ■, polycarbonate, o, PET (Mylar) ,•, and polyimide (Kapton), ▲, (reprinted from Jellink and Srinivasan 1984 with permission from the American Chemical Society)

A photoablation model based on the density of broken bonds resulting from absorbed photons has been advanced by Keyes et al. (1985) and seems to fit the observed ablation depth at high fluence better than Eq. (5.3.7). In this model, molecules are treated as polymers with a single type of bond. Assuming that the rate of bond breakage at depth x below the surface and at time t is proportional to the density of unbroken bonds, $n(x,t)$, and to the laser intensity, $I(x,t)$, one has

$$\frac{dn(x,t)}{dt} = -k_1 I(x,t) n(x,t). \tag{5.3.9}$$

Here, the decay of the intensity I with depth is modified from Beer's law to account for the moving interface or the fact that material is being ablated. Thus,

$$I(x,t) = I_0(t) \exp\left[-k_2 \int_0^x n(y,t)dy\right]. \tag{5.3.10}$$

Combining Eqs. (5.3.9) and (5.3.10) yields

$$\frac{dn(x,t)}{dt} = -k_1 n(x,t) I_0(t) \exp\left[-k_2 \int_0^x n(y,t)dy\right]. \tag{5.3.11}$$

The solution to Eq. (5.3.11), with initial condition

$$n(x,t)\,|_{t=0} = 0, \ x < 0$$

$$n(x,t)\,|_{t=0} = N, \ x \geq 0$$

and for constant intensity I_0 over time, is given by

$$\frac{n(x,t)}{N} = [1 + (\exp(k_1 I_0 t) - 1)\exp(-k_2 N x)]^{-1}. \tag{5.3.12}$$

The ablated depth (d) at time t may be calculated by equating the fraction of unbroken bonds in the left-hand side of Eq. (5.3.12) to some threshold F_T above which no ablation takes place and solving for x. This gives

$$x = d = \frac{1}{k_2 N} \ln[(\exp(k_1 I_0 t) - 1)/(\frac{1}{F_t} - 1)] \tag{5.3.13}$$

for $exp(k_1 I_0 t) > 1$. From this equation, it is seen that for small $I_0 t$ the depth d is proportional to $\ln(I_0 t)$ (here, it is assumed that enough time has passed for $\frac{n(x,t)}{N}$ to equal F_t so that $\ln(I_0 t)$ is positive) as in Eq. (5.3.7) based on Beer's law. On the other hand, for large $I_0 t$, the depth is proportional to $I_0 t$. This result may be an improvement on Eq. (5.3.7) in the sense that it may give a better fit to etch depth data when the fluence is not near to the threshold. From Eq. (5.3.12), one may also determine the time required in order to ablate a prespecified depth d. It is seen that this time is given by

$$t = \frac{1}{k_1 I_0} \ln[1 + \frac{1 - F_t}{F_t} \exp(k_2 N d)]. \tag{5.3.14}$$

This model has not been tested against etch depth data. However, its prediction of regions of logarithmic and linear increase of etch depth versus pulse fluence is in qualitative agreement with experimental data on excimer laser ablation of polyimide.

A model based on the concept of an absorbed photon flux threshold, Γ_T, below which photofragmentation is negligible and above which bond breakage can reach a critical density leading to polymer fragmentation and ejection from the bulk, has been proposed by Sutcliffe and Srinivasan (1986). Photon flux at time t and at depth x has been defined as

$$\Gamma(t,x) = I_0(t) \exp(-\alpha x)(\frac{\alpha \lambda}{hc}), \tag{5.3.15}$$

where h is Planck's constant, c is the speed of light, α is the absorption coefficient of the polymer, and λ is the wavelength. In addition, an effective concentration of absorbed photons above threshold at time t and at depth x is given by

$$\rho(t,x) = \int_0^t \theta(\Gamma(t',x) - \Gamma_T)dt',$$

(5.3.16a)

where

$$\theta(y) = y; \; y \geq 0$$

$$\theta(y) = 0; \; y < 0.$$

(5.3.16b)

Here, it is assumed that ablation takes place when $\rho(t,x)$ is equal or larger than a certain threshold, ρ_T. The numerical procedure used to evaluate the model in Eq. (5.3.16) involved dividing the polymer material into successive layers each of the same thickness x_j ($j = 1, 2, ...N$; $N = 10^3 - 10^4$) and dividing the pulse intensity curve over time into k subintervals t_i ($i = 1, 2, ..., k$) each of width Δt_i ($\simeq 20$ ps). Each subinterval represents a rectangular subarea or pulse that may be regarded as having a constant intensity. As such,

$$\rho(t_l, x_j) = \sum_{i=1}^l \theta \left[\Gamma(t_i, x_j) - \Gamma_T)\right] \Delta t_i.$$

(5.3.17)

Here, the sum is over more than a single pulse if l exceeds k. Ablation was assumed to commence in layer x_j when at any time t_l, $\rho(t_l, x_j) \geq \rho_T$. Two cases were considered when the layer reaches the ablation stage. Case 1 assumed that the ejected fragments have no absorption effect on the laser light. Case 2 assumed that the fragments absorbed the laser light with the same intensity α for the non-irradiated material. The etch depth was determined by the number of layers that reached the ablation condition. The model in Eq. (5.3.17) was fitted to etch depth measurements from experiments on PMMA and polyimide. Figure (5.3.2a) shows good fit to experimental observation in the low fluence range. The model predicts the non-linear behavior in the etch curve in that range, but failed to account for the plateau at higher fluence. The same type of fit was manifested also for etch results at 248 nm wavelength. It is noted also that the difference between Case 1 and Case 2 was negligible as seen in Fig. (5.3.2b). This result was attributable to the relatively low absorption coefficient of PMMA. In fact, Cases 1 and 2 were strikingly different for polyimide, a strong absorber. In the case of polyimide, model fit to experimental etch depth per pulse was good only near the threshold. It was interesting to note that the ablation condition ρ_T, as obtained from a fit of the model to experimental data, was the same for PMMA and polyimide in spite of the differences in their chemical and photochemical properties. In support of the model, it was argued that ρ_T would not be expected to change if ablation is caused by internal stress or pressure build-up, as postulated in the model, caused by the accumulation of fragments below the surface. As pointed out by the authors, this model brings out the importance of pulse width in pulsed-laser ablation and presents the new idea of an absorbed photon flux thresholds for fragmentation and for ablation. However, improvements in the model are still needed in order to obtain better predictions of etch characteristics.

A simple model taking into account the screening effect of the expelled fragments has been proposed by Lazare and Granier (1989). In this model, the instantaneous speed, v, with which the interface recedes into the material as a result of ablation is assumed to be proportional to the difference between the laser intensity at the surface and the ablation threshold. As such,

$$v(t) = k[I_0(t)(e^{-\beta(t)x(t)} - I_{th}(t)], \tag{5.3.18}$$

where $\beta(t)$ is the absorption coefficient of the expelled fragments at time t, k is the rate constant, $x(t)$ is the position of the interface at time t measured in relation to the original surface before the commence of ablation, $I_0(t)$ is the laser intensity $(\frac{MW}{cm^2})$ profile over the pulse duration, and $I_{th}(t)$ is the threshold of ablation at time t. Ablation occurs when the intensity reaching the surface exceeds the threshold I_{th}. Here, $v(t)$ was expressed in $\overset{0}{A}$ per $nsec$ and k in $\overset{0}{A}.$ $nsec^{-1}.$ $MW^{-1}.$ $cm^2.$

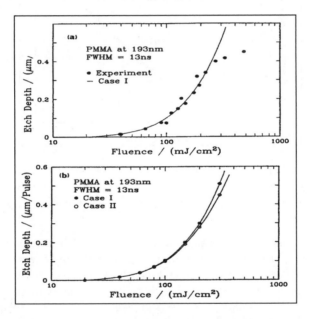

Fig. 5.3.2. Predicted and observed etch depth as a function of fluence for PMMA (reprinted from Sutcliffe and Srinivasan 1986 with permission from the American Institute of Physics. Experimental values from Srinivasan and Dreyfus 1985, Fig.1, with permission from Springer-Verlag)

The etch depth per pulse, d, is obtained by integrating the instantaneous velocity, $v(t)$, over the pulse duration $(0, t)$. In evaluating the integral, $k, \beta,$

and I_t were considered as constant. Also, only positive $v(t)$ in the pulse range contribute to the integral. In fitting the model to etch depth, one obtains an average value for β because in reality, β is a function of time depending on the expanding expelled fragments. The model was fitted to etch depth data as a function of fluence for PET in the case of 193 nm and 248 nm wavelengths. The fit was good up to one Joule per cm^2 for the 193 nm wavelength and up to $10^4 \frac{mj}{cm^2}$ for the 248 nm wavelength. Figure (5.3.3) shows the fit for 248 nm wavelength where the threshold, as measured experimentally using the quartz crystal microbalance, was $22 \frac{mj}{cm^2}$. From this fit, $\beta\ cm^{-1}$ was estimated to be 0.5×10^5 and k was 25. This model is similar to that by Sutcliffe and Srinivasan (1986) in the sense that ablation is dynamic and depends on the laser intensity profile or distribution above threshold over the time interval of the laser pulse.

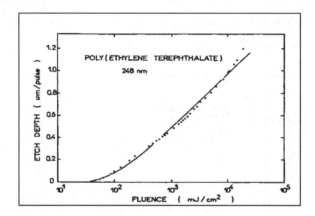

Fig. 5.3.3. Etch depth per pulse for PET and 248 nm wavelength. The solid curve is calculated from the model (reprinted from Lazare and Granier 1989 with permission from Taylor & Francis Ltd., http://www.tandf.co.uk/journals)

A photochemical model based on the number of broken bonds, $n(x,t)$, per unit volume at a distance x from the surface at time t was developed by Mahan et al. (1988) to predict the etch depth per pulse calculated over many pulses. The model considers the distribution of the laser intensity over the pulse duration and, similar to previous models, assumes that ablation occurs when the density of broken bonds exceeds a certain threshold value, n_{th}. The rate of change of $n(x,t)$ is expressed as

$$\frac{dn(x,t)}{dt} = \frac{p\alpha I(t)}{h v} \exp[-\alpha(x - s(t))].$$
(5.3.19)

Here, $s(t)$ is the position of the surface at the beginning of the pulse, p is the proportion of absorbed photons that contributes to bond breakage, $h v$ is the photon energy, α is the absorption coefficient, and $I(t)$ is the laser intensity at some time t during the pulse. The laser intensity in $\frac{J}{cm^2s}$ is expressed as,

$$I(t) = Fi(t), \qquad (5.3.20)$$

where F is the laser fluence in $\frac{J}{cm^2}$ and $i(t)$ is the normalized intensity distribution of the laser pulse over pulse duration. The pulse duration was considered over the interval $(-t_r, t_r)$ with maximum at $t = 0$. In this case, $\int_{-t_r}^{t_r} i(t)dt = 1$.

Integrating Eq. (5.3.19) with initial condition $n_{th}e^{-\alpha x}$ gives

$$n(x, t) = \frac{p\alpha F}{hv}e^{-\alpha x} \int_{-t_r}^{t} e^{\alpha s(t')}i(t')dt' + n_{th}e^{-\alpha x}. \qquad (5.3.21)$$

It is clear that the the position $s(t)$ of the ablation surface is obtained when $n(s, t) = n_{th}$. Replacing x by s in Eq. (5.3.21) and multiplying both sides by $\frac{e^{\alpha s}}{n_{th}}$ gives

$$J(t) = e^{\alpha s} = 1 + \frac{p\alpha F}{hvn_{th}} \int_{-t_r}^{t} J(t')i(t')dt'. \qquad (5.3.22)$$

Differentiating Eq.(5.3.22) with regard to t, gives

$$\frac{\partial(J(t))}{\partial t} = \frac{p\alpha F}{hvn_{th}}[i(t)J(t)], \qquad (5.3.23)$$

or

$$\frac{\partial(\ln J(t))}{\partial t} = \alpha\frac{ds}{dt} = \frac{p\alpha F}{hvn_{th}}[i(t)]. \qquad (5.3.24)$$

The integral of Eq. (5.3.24) over the interval $(-t_r, t_r)$ gives the etch depth, d, per pulse.

$$d = s(t_r) = \frac{pF}{hvn_{th}}. \qquad (5.3.25)$$

This equation showed good agreement with experimental data for PMMA and poly(α-methyl) styrene (PS) for fluence values between 0.1 and 0.5 $\frac{J}{cm^2}$, and for PI for fluence values between 0.1 and about 1.6 $\frac{J}{cm^2}$. As explained by the authors, the linear dependence of etch depth on fluence ceases to be linear for large fluence or for low fluence ($F < \frac{0.1J}{cm^2}$) in which case the present theory does not hold.

A simple surface photochemical etching model based on a single-step photochemical reaction has been considered by Bityurin (1999) within a general theoretical framework of multisteps (successive or branching) photochemical reactions accompanied by modification within the bulk of material. The interest in this study is thin polymer film modification by laser radiation below the ablation threshold. Consider the simple photochemical reaction

$$A \xrightarrow{hv} B, \qquad (5.3.26)$$

where A denotes an element of the initial material and B the reaction product. Based on Eq. (5.3.26), it is seen that the rate of change in the number concentration of A element (N_A) with time is

$$\frac{\partial N_A(z,t)}{\partial t} = -\eta_A \sigma_A N_A I(z,t), \qquad (5.3.27)$$

where η_A is the quantum yield, σ_A is the absorption cross-section, and I is the light intensity. Also, the rate of change of the light intensity with depth is

$$\frac{\partial I(z,t)}{\partial z} = -(\sigma_A N_A + \sigma_B N_B)I(z,t). \qquad (5.3.28)$$

Here, $N_A + N_B = N_0$. The initial and boundary conditions are

$$N_A(z,0) = N_0; \quad N_B(z,0) = 0; \quad I(0,t) = I_0(t), \qquad (5.3.29)$$

where $I_0(t)$ is a function of time which is determined by the light source.

The surface ablation velocity (V_a) is assumed to depend linearly on N_A and N_B through the expression

$$V_a = \frac{dz_s}{dt} = (G_A N_A + G_B N_B)I_0(t). \qquad (5.3.30)$$

Equations (5.3.27)-(5.3.30) were expressed in dimensional form and solved to give the etch depth as a function of exposure dose. Comparisons to PMMA data ablated by a $0.5 \frac{mJ}{cm^2}$ Nd:YAP (216 nm) laser fluence shows that, for a film thickness of 600 nm, the model was adequate in explaining the ablation depth up to 25 $\frac{J}{cm^2}$ of exposure dose for etching in vacuum (5.10^{-3} Torr) and 15 $\frac{J}{cm^2}$ in air. However, the model was not adequate in explaining experimental data for a thinner film of about 400 nm. Thus, the model does not explain the dependence of the etch rate on initial film thickness.

A bulk model, similar in approach to the above model, gave better fit to the experimental data (Bityurin et al. 1997). In this model, the photochemical reaction in vacuum is expressed as

$$A \overset{h\nu}{\rightarrow} B + P_B, \qquad (5.3.31)$$

where P_B designates the volatile reaction products that diffuse through the polymer and leave the surface. For PMMA, P_B represents the ester group. In the case of polymer irradiation in air, there are two reaction steps involved, namely the production of the ester group in the first step and the subsequent elimination of oxidation products in the second step. This reaction scheme can be expressed as

$$A \overset{h\nu}{\rightarrow} B + P_B \overset{h\nu}{\rightarrow} C + P_C. \qquad (5.3.32)$$

Here, B and C are products that remain in the polymer film while P_B and P_C are the volatile products that leave the film. The governing equations in one dimension for the two-step chemical reaction, in Lagrange representation where a species concentration is based on the initial volume without regard to material contraction, can be expressed as

$$\frac{\partial A_L}{\partial t} = -\eta_A \sigma_A A_L I; \quad \frac{\partial B_L}{\partial t} = \eta_A \sigma_A A_L I - \eta_B \sigma_B B_L I, \quad (5.3.33)$$

and

$$\frac{\partial P_B}{\partial t} = \eta_A \sigma_A A_L I; \quad \frac{\partial P_C}{\partial t} = \eta_B \sigma_B B_L I. \quad (5.3.34)$$

Here, $C_L = A_0 - A_L - B_L$. Assuming no significant effect of light reflection at the substrate-film interface, the change in light intensity with depth can be expressed as

$$\frac{\partial I}{\partial Z_L} = -\alpha_L I = -(\sigma_A A_L + \sigma_B B_L + \sigma_C C_L)I, \quad (5.3.35)$$

where α_L is the absorption coefficient. In terms of the Lagrange coordinates, the interface between film and substrate corresponds to $Z_L = h_{init}$ (the initial film thickness), and $Z_L = 0$ corresponds to the film surface. These positions are time-independent. The Lagrange coordinates correspond to the Euler coordinates at $t = 0$. The initial and boundary conditions for Eqs. (5.3.33)-(5.3.35) are

$$A_L(Z_L, 0) = A_0; \quad B_L(Z_L, 0) = C_L(Z_L, 0) \quad (5.3.36)$$
$$= P_B(Z_L, 0) = P_C(Z_L, 0) = 0;$$
$$I(0, t) = I_0(t).$$

From the solution to Eqs. (5.3.33)-(5.3.35), one can determine the solution in terms of the Euler coordinates using the following transformations:

$$Z = Z_L + \int_{Z_L}^{h_{init}} (P_B(Z_L, t)v_B + P_C(Z_L, t)v_C)dz'_L, \quad (5.3.37)$$

$$A_E = \frac{A_L}{1 - P_B v_B - P_C v_C}; \quad B_E = \frac{B_L}{1 - P_B v_B - P_C v_C}, \quad (5.3.38)$$

where v_B and v_C correspond to the specific free volume of P_B and P_C, respectively. A_E and B_E are the Euler concentrations of species A and B, and $(1 - P_B v_B - P_C v_C)$ represents contraction of the polymer material. The current film thickness, $h(t)$, can be obtained by setting $Z_L = 0$ in Eq. (5.3.37) to obtain

$$h(t) = h_{init} - \int_0^{h_{init}} (P_B(Z_L, t)v_B + P_C(Z_L, t)v_C)dz'_L. \quad (5.3.39)$$

Equations (5.3.33)-(5.3.38) may be represented in dimensionless form. For dimensionless time, the quantity τ is defined as

$$\tau = \int_0^t \eta_A \sigma_A I_0(t) dt' = \frac{dose}{D_{CH}}, \tag{5.3.40}$$

where $D_{CH} = \frac{h\nu}{\eta_A \sigma_A}$. This implies that the dose in $\frac{J}{cm^2}$ is equal to $h\nu \int_0^t I_0(t) dt'$. Other dimensionless parameters are

$$\mu = \frac{\eta_B \sigma_B}{\eta_A \sigma_A}; \quad \beta_B = \frac{\sigma_B}{\sigma_A}; \quad \beta_C = \frac{\sigma_C}{\sigma_A}; \quad \varepsilon_B = A_0 v_B; \quad \varepsilon_C = A_0 v_C. \tag{5.3.41}$$

From the solution to the model Eqs. (5.3.33)-(5.3.38) one may obtain $h(t)$ in Eq. (5.3.39) as a function of t or the exposure dose. The solid curves in Figs. (5.3.4a,b) represent film thickness, $h(t)$, as a function of exposure dose fitted to experimental data. It is seen that the fit is good for both figures except perhaps line 1 in Fig. (5.3.4a). This discrepancy between model and observations was attributed by the authors to the influence of light reflected from the interface between film and substrate, which was not considered in the model. The parameters used to obtain the model fits to data were $\mu = 0.16$; $\beta_B = 15$; $\beta_C = 4$; $\varepsilon_B = 0.4$; $\varepsilon_C = 0.5$; and $D_{CH} = 1.754 \frac{J}{cm^2}$ for Fig. (5.3.4a). For Fig. (5.3.4b), pertinent parameters were $\mu = 0$; $\beta_B = 20$; $\varepsilon_B = 0.4$; and $D_{CH} = 1.754 \frac{J}{cm^2}$. In both figures, thin films of PMMA on a silicon substrate were exposed to a subthreshold laser fluence of 0.5 $\frac{mJ}{cm^2}$.

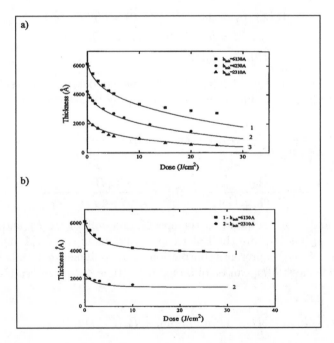

Fig. 5.3.4. Film thickness as a function of exposure dose (**a**) in air and (**b**) in vacuum (reprinted from Bityurin et al. 1997 with permission of the SPIE)

5.4 Photophysical and Other Models

A model combining photothermal and photochemical contributions to etching has been proposed by Srinivasan et al. (1986) to explain etch depth data for relatively high (larger than $1 J/cm^2$) excimer laser fluence. The photochemical mechanism was based on Beer's law and the photothermal on the Arrhenius equation. As such, the etch depth per pulse is given by

$$d = \frac{1}{\alpha} \ln(\frac{F_0}{F_{th}}) + A \exp(-\frac{E_a}{kT}). \qquad (5.4.1)$$

Temperature in the polymer sample arising from photon absorption is directly proportional to fluence. This may be expressed as

$$F - F_0 \propto l_p T, \qquad (5.4.2)$$

where l_p is the depth of penetration of the absorbed photon and is given by the first expression on the right-hand side of Eq.(5.4.1). Hence, it is seen that

$$T \propto (F - F_0)/\frac{1}{\alpha} \ln(\frac{F}{F_{th}}). \qquad (5.4.3)$$

Here, F_0 is of order F_{th} and can be neglected when F is large (in $\frac{J}{cm^2}$). Substituting Eq. (5.4.3) into Eq. (5.4.2) gives

$$d = \frac{1}{\alpha} \ln(\frac{F_0}{F_{th}}) + A \exp\left[\frac{-E}{\alpha(F - F_0)} \ln(\frac{F}{F_{th}})\right]. \qquad (5.4.4)$$

Here, E contains k, E_a, and the proportionality constant.

From Eq. (5.4.4), it is seen that the etch depth $d = d_p + d_T$ (sum of the photochemical and thermal components). As such,

$$\ln(d_T) = \ln A - \frac{E}{\alpha(F - F_0)} \ln(\frac{F}{F_{th}}), \qquad (5.4.5)$$

and

$$d_p = \frac{1}{\alpha} \ln(\frac{F_0}{F_{th}}). \qquad (5.4.6)$$

For large fluence, d_T was evaluated as the difference between the observed etch depth d and d_p. Plots of $\ln(d_T)$ versus $\ln(\frac{F}{F_{th}})/F$ for plyimide, PMMA, and TNS2 photoresist showed a negative linear trend thus confirming the validity of Eq. (5.4.5). As explained by the authors, Eq. (5.4.4) suggests that for low fluences, the etch depth is controlled by the photochemical component which gives rise to a logarithmic relationship between etch depth and fluence. At higher fluences, the thermal component dominates and gives rise to an increase in etch depth with a linear relationship between etch depth and fluence. At still higher fluences, the thermal component reaches the limit $\ln A$ beyond which the increase in etch depth with fluence is affected by the photochemical component. As a result, the slope of the etch rate curve decreases causing the curve to go through an inflection point. The prediction of the three regions depicting etch rate as a function of fluence is in general agreement with experimental observations. Figure (5.4.1) for TNS2 shows the three regions of the curve as explained above. It shows also agreement between the model in Eq. (5.4.4) and observed etch depth per pulse for 193 nm and 248 nm excimer lasers at high fluence. Similar results were obtained for polyimide and PMMA.

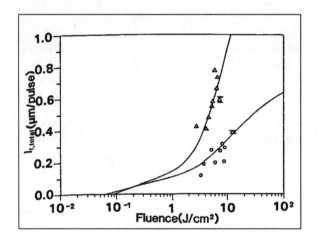

Fig. 5.4.1. Etch depth per pulse as a function of fluence for TNS2 in the case of 193 nm (o) and 248 nm ($\triangle$) excimer laser. Solid line is from Eq. (5.4.4) with d = l$_{f,\,total}$ on the y-axis (reprinted from Srinivasan et al. 1986 with permission from the American Institute of Physics)

A model for photoablation based on single and two photon absorption has been considered by Sauerbrey and Pettit (1989) where the ablation depth per pulse is determined by the penetration depth where the photon density reaches a certain threshold. The governing equations for the density of the light absorbing chromophores, $\rho = \rho(x, t)$, and the incident photon flux, $I = I(x, t)$, are given by

$$\frac{\partial \rho}{\partial t} = \sigma_1(\rho_0 - \rho)I + \sigma_2(\rho_0 - \rho)I^2, \tag{5.4.7}$$

$$\frac{\partial I}{\partial x} = -\sigma_1(\rho_0 - \rho)I - \sigma_{1p}\rho I - \sigma_2(\rho_0 - \rho)I^2, \tag{5.4.8}$$

where ρ_0 is the density of the chromophores at time $t = 0$, x is the depth as measured from the initial material surface, σ_1 and σ_2 are the absorption cross-sections for one and two photon absorption, respectively, and σ_{1p} is the absorption cross-section for the plume assuming only single photon absorption. Also, it is seen that $I(0, t) = I(t)$ and $\rho(x, 0) = 0$. Integrating Eq. (5.4.7), one obtains

$$\rho(x,t) = \rho_0(1 - \exp(-\int_0^t \sigma_1 I(x,t') + \sigma_2 I^2(x,t')dt'. \tag{5.4.9}$$

Ignoring absorption by the plume ($\sigma_{1p} = 0$) and substituting Eq. (5.4.9) in Eq. (5.4.8), it is seen that

$$\frac{\partial I}{\partial x} = -\rho_0[\sigma_1 I(x,t) + \sigma_2 I^2(x,t)]\exp(-\int_0^t \sigma_1 I(x,t') + \sigma_2 I^2(x,t')dt'. \tag{5.4.10}$$

Since the pulse duration is much shorter than the time interval between pulses, one may integrate Eq. (5.4.10) from zero to infinity to obtain the photon density $S(x)$ as a function of depth. For single photon absorption ($\sigma_2 = 0$), this integral is seen to give

$$\frac{dS}{dx} = -\rho_0(1 - e^{-\sigma_1 S}), \tag{5.4.11}$$

where $S = S(x) = \int_0^\infty I(x,t)dt$. Here, the fluence F is equal to $hvS(x)$. It is readily seen from Eq. (5.4.11) that using the Taylor series approximation for small $\sigma_1 S$ (close to threshold) gives

$$\frac{dS}{dx} = -\sigma_1 \rho_0 S = -\alpha S. \tag{5.4.12}$$

It is seen that the solution to Eq.(5.4.12), with $F = hvS$, gives the etch depth per pulse based on Beer's law as in Eq. (5.4.6). In general, the etch depth per pulse is obtained by integrating Eq. (5.4.11) from $S_0 = S(0)$ to some threshold photon density S_{th}. This gives

$$d = x = \frac{1}{\rho_0}(S_0 - S_{th}) + \frac{1}{\alpha}\ln\left(\frac{1 - e^{-\sigma_1 S_0}}{1 - e^{-\sigma_1 S_{th}}}\right). \tag{5.4.13}$$

Equation (5.4.13) shows that for large fluence, etch depth becomes linearly dependent on F.

If single photon absorption by the plume is considered, it is seen from the above derivations that the depth per pulse becomes

$$\frac{\partial I}{\partial x} = -\rho_0[\sigma_1 I(x,t)\exp(-\int_0^t \sigma_1 I(x,t')dt')$$
$$-\frac{\sigma_{1p}}{\sigma_1}\sigma_1 I(x,t)\rho_0(1 - \exp(-\int_0^t \sigma_1 I(x,t')dt'))]. \tag{5.4.14}$$

Integrating Eq. (5.4.14) from S_0 to S_{th}, yields

$$d_p = \frac{1}{\rho_0}\int_{S_{th}}^{S_0}[\sigma_{1p}S + (1 - \frac{\sigma_{1p}}{\sigma_1})(1 - e^{-\sigma_1 S})]^{-1}dS. \tag{5.4.15}$$

For multiphoton absorption without plume attenuation (Pettit and Sauerbrey 1993), it is seen that Eq. (5.4.8) may be expressed in general as

$$\frac{\partial I(x,t)}{\partial x} = -\sum_{n=1}^{\infty} n\sigma_n(\rho_0 - \rho)I^n. \tag{5.4.16}$$

Here, the factor n reflects the fact that each absorption reduces the laser flux by n photons. If the predominant ablation mechanism involves the absorption of n photons, then Eq. (5.4.16) reduces to

$$\frac{\partial I(x,t)}{\partial x} = -n\sigma_n(\rho_0 - \rho)I^n. \tag{5.4.17}$$

Also,

$$\frac{\partial \rho}{\partial t} = \sigma_n(\rho_0 - \rho)I^n. \tag{5.4.18}$$

Solving Eq. (5.4.18) for ρ and substituting the solution into Eq. (5.4.17) yields

$$\frac{\partial I}{\partial x} = -n\sigma_n I^n(x,t)\rho_0 \exp(-\int_0^t \sigma_n I^n(x,t')dt'). \tag{5.4.19}$$

Integrating Eq. (5.4.19) from zero to infinity gives the photon density equation

$$\frac{dS}{dx} = -n\rho_0[1 - \exp(-\int_0^{\infty} \sigma_n I^n(x,t')dt')]. \tag{5.4.20}$$

Let $\int_0^{\infty} I^n(x,t)dt = K_n \left[\int_0^{\infty} I(x,t)dt\right]^n = K_n S^n$. Here, $K_1 = 1$ (single photon case) and $K_n = \frac{A_n}{\tau^{n-1}}$, where A_n is a constant dependent on the pulse shape, and τ is a laser pulse characteristic time. For instance when $n = 2$, Pettit and Sauerbrey (1993) showed that $K_2 = \frac{1}{[t_1(x) - t_2(x)]}$. For $n = 2$, it is seen , from the mean value theorem in integral calculus, that $\int_0^{\infty} I^2(x,t)dt = I(x,\tilde{t}(x))\int_0^{\infty} I(x,t)dt$ for some value $\tilde{t}$. Also, $\int_0^{\infty} I(x,t)dt = [t_1(x) - t_2(x)]I(x,\tilde{t}(x))$. Hence, $\int_0^{\infty} I^2(x,t)dt = \frac{1}{[t_1(x)-t_2(x)]}\left[\int_0^{\infty} I(x,t)dt\right]^2 = K_2\left[\int_0^{\infty} I(x,t)dt\right]^2$ As such, Eq.(5.4.20) reduces to

$$\frac{dS}{dx} = -n\rho_0[1 - \exp(-\sigma_n K_n S^n)]. \tag{5.4.21}$$

Integrating Eq. (5.4.21) from S_0 to S_{th} yields the etch depth per pulse

$$d_n = \frac{1}{n\rho_0}\int_{S_{th}}^{S_0} [1 - \exp(-\sigma_n K_n S^n)]^{-1}dS. \tag{5.4.22}$$

Equation (5.4.13) was fitted to polyimide etch depth data as shown in Fig. (5.4.2). The model is in excellent agreement with the observed data. It shows also that pulse length on a time scale in nanoseconds has no effect on etch depth. Single photon absorption ignoring plume attenuation seems to be a good model for polyimide ablation because polyimide has a high absorption coefficient, a small etch depth per pulse, and plume absorption similar to the bulk material. Where photon absorption is relatively low as for PMMA, ablation by nanosecond excimer laser generates large decomposition products which have an effect on photon absorption. As such, plume absorption is a significant factor. In this case, it is shown that an excellent fit to the etch depth data was obtained using Eq. (5.4.15), Fig. (5.4.3). On the other hand, two photon absorption or single photon absorption, Eqs. (5.4.22) and (5.4.13), did not give adequate fit. For very short pulse length (in femtoseconds), the plume effect is not expected to be significant since it occurs after the laser pulse (Garrison and Srinivasan 1984). Multiphoton absorption is also likely to dominate under high intensity femtosecond laser pulses. Such is the case for Teflon ablation with KrF excimer laser. Figure(5.4.4) shows an excellent fit of the model in Eq. (5.4.22) for two photon absorption ($n = 2$), solid curve. The single photon absorption model, Eq.(5.4.13), dashed curve, did not show as good a fit.

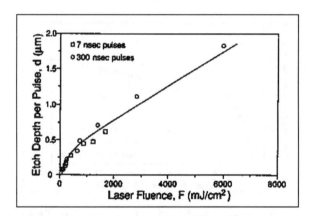

Fig. 5.4.2. Etch depth as a function of fluence for polyimide in the case of XeCl laser (308 nm) with 7 ns ($\square$) and 300 ns (O) pulse lengths. Data are from Taylor et al. 1987 and the solid curve is from Eq. (5.4.13) with $\sigma_1 = 10^{-18} \text{cm}^2$ and $\alpha = 6.5$ μm^{-1} (reprinted from Sauerbrey and Pettit 1989 with permission from the American Institute of Physics)

The above photon absorption model ignored the transition or return of a chromophore from an excited state to the ground state. This is justifiable if the transition time is large relative to the pulse duration time. Transition to the ground state is expected to become significant if the transition or decay time is small relative to the pulse duration. For such a situation, Eq. (5.4.7) in the case of a single photon absorption may be expressed as (Pettit and Sauerbrey 1993)

$$\frac{\partial \rho}{\partial t} = \sigma_1 (\rho_0 - \rho) I - \frac{\rho}{\tau}, \tag{5.4.23}$$

where τ is the transition or decay time from an excited state to the ground state. If $E(x,t)$ represents the absorbed energy per unit volume, then the ablation depth is determined by the x value where $E(x,\infty)$ reaches a certain threshold value, E_{th}. The rate equation for $E = E(x,t)$ is

$$\frac{\partial E}{\partial t} = \sigma_1 \rho_0 I \, h\upsilon. \qquad (5.4.24)$$

The coupled system of Eqs. (5.4.24), (5.4.23), and (5.4.8), with $\sigma_2 = \sigma_{1p} = 0$, was solved numerically to obtain the absorbed energy $E(x,t)$ and the etch depth for polyimide ablation by ArF and KrF excimer lasers with a pulse duration of 15 ns. Etch depth from the model showed good agreement with experimental observations except at high fluence which was attributed to light absorption by the plume.

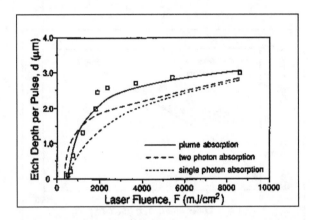

Fig. 5.4.3. Etch depth per pulse for PMMA as a function of fluence for a KrF laser with 20 ns pulse duration. The solid line is calculated from Eq. (5.4.15) for $\rho_0 = 2.7 \times 10^{21} \text{cm}^{-3}$, $\sigma_1 = 4 \times 10^{-19} \text{cm}^2$, and $\sigma_{1p} = 10^{-17} \text{cm}^2$. The dashed line is from two photon absorption, Eq. (5.4.22), for $F_{th} = 350 \, \frac{\text{mJ}}{\text{cm}^2}$, $\sigma_2 K_2 = 1.2 \times 10^{-37} \text{cm}^4$, and $\rho_0 = 9 \times 10^{22} \text{cm}^{-3}$. The dotted line is from Eq. (5.4.13) for single photon absorption with $F_{th} = 500 \, \frac{\text{mJ}}{\text{cm}^2}$, $\sigma_1 = 10^{-25} \text{cm}^2$, and $\alpha = 1 \, \mu\text{m}^{-1}$.
Experimental values from Keuper and Stuke 1987 with permission from Springer-Verlag (reprinted from Sauerbrey and Pettit 1989 with permission from the American Institute of Physics)

A photophysical laser ablation model which considers a single photon chromophore absorption, giving rise to an excited state and a ground state, has been discussed by Luk'yanchuk et al. (1993a). The model combined thermal and non-thermal ablation mechanisms. It differs from previous models in that it allows for non-thermal ablation through desorption of excited species. Considering the ground and excited states, the number densities of chromophores in the excited (N^*) state and in the ground (N_0) state can be expressed as

$$\frac{\partial N^*}{\partial t} = v\frac{\partial N^*}{\partial z} + \frac{\sigma I}{h\nu}(N_0 - N^*) - \frac{N^*}{t_T}, \qquad (5.4.25)$$

$$\frac{\partial N_0}{\partial t} = v\frac{\partial N_0}{\partial z} - \frac{\sigma I}{h\nu}(N_0 - N^*) + \frac{N^*}{t_T}, \qquad (5.4.26)$$

where t_T is the thermal relaxation time and σ is the absorption cross-section. The ablation velocity caused by desorption of ground and excited species is expressed as

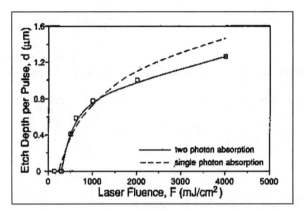

Fig. 5.4.4. Etch depth per pulse in Teflon as a function of fluence for a KrF laser with a laser pulse of 300 fs in length. The solid curve is calculated from Eq. (5.4.22) for two photon absorption with $F_{th}=300 \frac{mJ}{cm^2}$, $\sigma_2K_2 =3\times10^{-35}cm^4$, and $\rho_0=9\times10^{21}cm^{-3}$. The dashed line is caculated from Eq. (5.4.13) for single photon absorption with $F_{th}=250 \frac{mJ}{cm^2}$, $\sigma_1=10^{-25} cm^2$, and $\alpha=1.9\ \mu m^{-1}$. Experimental data from Kueper and Stuke 1989 (reprinted from Sauerbrey and Pettit 1989 with permission from the American Institute of Physics)

$$v = v_A \exp(-\frac{\Delta E}{T_s})(\frac{N_{0s}}{N}) + v_{A^*} \exp(-\frac{\Delta E^*}{T_s})(\frac{N_s^*}{N}), \qquad (5.4.27)$$

where $N = N_0 + N^*$, $\Delta E's$ are the activation energies, and the subscript s refers to the surface. Here, the ablation depth per pulse is defined as

$$\Delta h = \int_0^\infty \bar{v}(t)dt, \qquad (5.4.28)$$

where $\bar{v}(t) = v - v[T_s = T\infty]$. The integral accounts for ablation during the transition state, during the stationary ablation state, and after the laser pulse because of stored energy within the irradiated material. Additionally, the rate of change of the light intensity with depth can be described by

$$\frac{\partial I}{\partial z} = -\sigma(N_0 - N^*)I, \tag{5.4.29}$$

and the temperature at the solid surface by

$$\frac{\partial T}{\partial t} = v\frac{\partial T}{\partial z} + D_T\frac{\partial^2 T}{\partial z^2} + \frac{D_T}{\kappa}Q, \tag{5.4.30}$$

where D_T and κ are the thermal diffusivity and conductivity, respectively and $Q = \frac{h v N_{A^*}}{t_T}$ is the heat source. The heat flux at the surface $(z = 0)$ is given by

$$k\frac{\partial T}{\partial z} \mid_{z=0} = \varrho[\frac{\Delta E}{M}v_A \exp(-\frac{\Delta E}{T_s})(\frac{N_{0s}}{N}) + \frac{\Delta E^*}{M}v_{A^*} \exp(-\frac{\Delta E^*}{T_s})(\frac{N_s^*}{N})]. \tag{5.4.31}$$

Here, ϱ is the material density, M is the average mass of ablated material, and the enthalpy is $\frac{\Delta E}{M}$. The boundary conditions employed were

$$N_0(z \to \infty) = N, \quad N^*(z \to \infty) = 0, \quad T(z \to \infty) = T(\infty),$$

$$I(z = 0) = I_s(t). \tag{5.4.32}$$

The initial conditions were

$$N_0(t = 0) = N, \quad N^*(t = 0) = 0, \quad T(t = 0) = T(\infty). \tag{5.4.33}$$

For model calculations, N was taken to be $6 \times 10^{21} cm^{-3}$, $\Delta E = 3\ eV$ to $5.0\ eV$ (order of the bond breaking energy) and $\Delta E^* = 0.3\ eV$. The velocities v_A and v_{A^*} were of the order of the sound velocities and were chosen to be 3×10^5 cm/s and 10^5cm/s, respectively. In Eq. (5.4.30), $\frac{\Delta E}{M} = 10^3\ J/g$ and $\frac{\Delta E^*}{M} = 10^2\ J/g$. Model calculations based on stationary solutions revealed the following salient results concerning the effect of the relaxation time t_T.
1. The ablation velocity increased with an increase in the relaxation time t_T.
2. The surface temperature rise decreased with an increase in the relaxation time t_T. Also, the concentration of excited species on the surface increased significantly with an increase in t_T. These conditions can explain the occurrence of fairly high ablation rates at moderate surface temperatures (caused mainly by desorption of excited species with low activation energy). For instance, at relaxation times $t_T > 10^{-9}\ s$, the model predicts ablation at surface temperatures of about $2000^0 C$. This result is in contrast to the predictions of very high surface temperatures of about $6000\ K$ or higher for fluences near the ablation threshold based on a purely thermal model (Cain et al. 1992b).

Further analysis of this model concerning the stability of the ablation front (Luk'yanchuk et al. 1993b) revealed that thermal instabilities are suppressed for relaxation times $t_T \geq 10^{-10}$s. On the other hand, thermal instability occurs in metals where $t_T < 10^{-11}$s. Barring other instability effects (such as internal stress, non-uniform laser intensity), these thermal instability results can explain why laser ablation gives rough surfaces in metals and smooth surfaces in polymers.

Luk'yanchuk et al. (1996a) re-examined the above model by considering four states (two photon absorption) and the unsteady-state situation. Figure (5.4.5) gives a schematic diagram of the electronic transitions and relaxation pathways considered in the model. Here, excitation proceeds from N_0 to N_1 and N^* to N_2. State N^* may be regarded as more stable than N_1. Inclusion of this state allows for exclusion of $N_1 \to N_0$ transition through stimulated emission. In this model, the relaxation times τ_1 and τ_2 are assumed to be fast and much smaller than the thermal relaxation time, t_T, and the excitation time $\tau_{ex} = \frac{h\nu}{I\sigma}$. On the other hand, the relaxation times τ_{10} and τ_{20} are assumed to be slow and much larger than t_T and the excitation time, τ_{ex}. Analogous to the above photophysical model, the governing equations for the concentration of excited chromophores N^*, the laser intensity I, and the temperature T are expressed as

$$\frac{\partial N^*}{\partial t} = v\frac{\partial N^*}{\partial z} + \frac{I\sigma_{01}}{h\nu}(N - N^*) - \frac{N^*}{t_T}, \qquad (5.4.34)$$

$$\frac{\partial I}{\partial z} = -I\sigma_{01}[N - N^* + sN^*], \qquad (5.4.35)$$

$$\frac{\partial T}{\partial t} = v\frac{\partial T}{\partial z} + D_T\frac{\partial^2 T}{\partial z^2} + (I\sigma_{01}s + \frac{h\nu}{t_T})\frac{N^*}{\rho c_p}. \qquad (5.4.36)$$

Here $s = \frac{\sigma_{12}}{\sigma_{01}}$, and N is the total number density of chromophores.

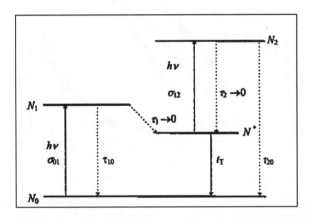

Fig. 5.4.5. Schematic diagram of the four states with transition and relaxation pathways considered in the model (reprinted from Luk'yanchuk st al. 1996a with permission from Springer-Verlag)

The ablation velocity and the heat flux at the surface ($z = 0$) are given by Eqs. (5.4.27) and (5.4.31) where $\frac{N_{0s}}{N}$ is replaced by $\frac{N-N^*}{N}$. The boundary conditions are

$$N^*(z \to \infty) = 0, \ T(z \to \infty) = T_\infty,$$

$$I(z = 0) = I_0(t) \exp\left[-\alpha_g \int_0^t v(t')dt',\right] \quad (5.4.37)$$

where α_g is the absorption coefficient of the plume. The initial conditions are

$$N^*(t = 0) = 0, \ T(t = 0) = T_\infty. \quad (5.4.38)$$

Numerical solution of the model equations using the finite element method was obtained, and predicted ablated depth per pulse as a function of fluence was compared to experimental data for polyimide at 193, 248, 308, and 351 wavelengths, Fig. (5.4.6). The solid lines showing good fit to the data are from the numerical solution of the governing equations using parameters listed in Table 5.4.1. The dashed curves show almost as good a fit and are obtained from least square fit of the formula

$$F = B \exp(\beta \Delta h) \ln^{-1}(\frac{A}{\Delta h}). \quad (5.4.39)$$

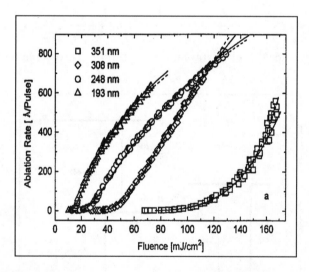

Fig. 5.4.6. Ablation depth per pulse as a function of fluence for polyimide with different wavelengths. Experimental data from Kueper et al. 1993 (reprinted from Luk'yanchuk et al. 1996a with permission from Springer-Verlag)

This equation is obtained by assuming that the maximum surface temperature $T_{s\ Max}$ is proportional to $F \exp(-\beta \Delta h)$ and substituting in the equation (Luk'yanchuk et al. 1994)

$$\Delta h = A \exp[-\frac{\Delta E}{T_{s\ Max}}]. \quad (5.4.40)$$

It should be noted that Eq. (5.4.40), and therefore Eq. (5.4.39), is based on a purely thermal model of ablation (Arrhenius type) and holds for low fluences.

The main feature of these photophysical models is their ability to explain (1) fairly high ablation rates at moderate surface temperatures, which cannot be explained based on a purely thermal model; (2) activation energies smaller than bond-breaking energies caused by the desorption of excited species; (3) surface stability for polymer ablation; and (4) Arrhenius-type behavior near the ablation threshold.

Table 5.4.1. Parameters used in the numerical calculations of the model (reprinted from Luk'yanchuk et al. 1996a with permission from Springer-Verlag)

	$\lambda = 193$ nm	$\lambda = 248$ nm	$\lambda = 308$ nm	$\lambda = 351$ nm
ΔE, eV	3	3	3	3
ΔE^*, eV	1.9	1.27	1.47	1.54
v_A, cm/s	10^6	10^6	10^6	10^6
v_A^*, cm/s	6.5×10^7	5.7×10^5	8.3×10^5	5.36×10^6
ΔH, J/g	2.3×10^3	2.3×10^3	2.3×10^3	2.3×10^3
ΔH^*, J/g	0	7.4×10^2	8.5×10^2	4.3×10^2
$\alpha_0 = \sigma_{01} N$, cm^{-1}	3.4×10^5	3×10^5	8.6×10^4	3.6×10^4
α_g, cm^{-1}	1.7×10^5	1.56×10^5	4.3×10^4	0
t_T, ps	18200	528	800	179
s	20	0	17	21.7
c_p, J/g K	1.1	1.1	1.1	1.1
ρ, g/cm^3	1.42	1.42	1.42	1.42
D_T, cm^2/s	0.001	0.001	0.001	0.001
N, cm^{-3}	6×10^{21}	6×10^{21}	6×10^{21}	6×10^{21}
T_∞, K	300	300	300	300

The above photophysical models assume that ablation takes place on the surface within a thin layer substantially smaller than the absorption depth of the laser intensity. Further analysis considered stresses induced within the bulk and their influence on the ablation rate (Luk'yanchuk et al. 1996b). The effect of stress (σ) on the activation energies of excited species and of ground state is described in terms of the linear equations

$$\Delta E(\sigma) = \Delta E_0 - \gamma\sigma, \qquad (5.4.41)$$

$$\Delta E^*(\sigma) = \Delta E_0^* - \gamma^*\sigma, \qquad (5.4.42)$$

where γ is a constant, and $\gamma^* > \gamma$. For PMMA, $\gamma = 1.7 \times 10^{-21}$ cm^3. The interest here is in the formation and desorption of light defects or fragments (such as CO, CN, and CH_2) generated from bond breaking. The number density of these light fragments (or gases) generated within the bulk can be described by the diffusion equation

$$\frac{\partial N_1}{\partial t} = v \frac{\partial N_1}{\partial z} + \frac{\partial}{\partial z}(D_1 \frac{\partial N_1}{\partial z}) + Q_1. \tag{5.4.43}$$

The source term is given by

$$Q_1 = (N_{01} - N_1)\left(\eta \frac{\sigma_{01} I}{h v} + k_1 \exp(-\frac{\Delta E_1(\sigma_z)}{k_B T}) + k_1^* \exp(-\frac{\Delta E_1^*(\sigma_z)}{k_B T})\right), \tag{5.4.44}$$

where the first term is the contribution from photochemical bond breaking with quantum yield η and the other terms are contributions from the ground state and the excited state molecules, respectively. Mass loss caused by surface ablation and by the flux of light fragments or gases leaving the surface are described by

$$\Delta M = \rho \int_0^\infty v(t)dt + \int_0^\infty -m_1 J_1 dt. \tag{5.4.45}$$

Here, m_1 is the average mass of the light fragments, and J_1 represents their flux at the surface which is given by

$$-J_1 = D_1 \frac{\partial N_1}{\partial z} \Big|_{z=0} = \beta N_1 \Big|_{z=0}. \tag{5.4.46}$$

Removal of light material becomes important near the threshold where it can exceed removal by ablation. Equation (5.4.46) is a boundary condition for Eq. (5.4.43). The other boundary condition is $N_1 \Big|_{z \to \infty} = 0$. An initial condition is $N_1(t = 0) = 0$. Since the activation energies in Eq.(5.4.44) are functions of stress, a coupling equation for stress must be employed. This is given by the thermoelasticity equation (Landau et al. 1986)

$$\frac{\partial^2 \sigma_z}{\partial t^2} = c_s^2 \frac{\partial^2 (\sigma_z - p)}{\partial z^2} - \frac{E_Y}{3(1 - 2\mu)} \frac{\partial^2}{\partial t^2}[\alpha_v(T - T_0)], \tag{5.4.47}$$

where c_s is the sound velocity, E_Y is Young's modulus of elasticity, μ is the Poisson ratio, and α_v is the volume thermal expansion coefficient.. The boundary conditions are

$$\sigma_z(t, z = 0) = p(t, z = 0), \ \sigma_z(t, z \to \infty) = 0. \tag{5.4.48}$$

The photophysical ablation process with stress is now described by Eqs. (5.4.34)-(5.4.38) and Eqs.(5.4.41)-(5.4.48). It is seen from Eqs.(5.4.41) and (5.4.42) that stress reduces the activation energy and can facilitate photophysical ablation. Model solution for the concentration of defects or light fragments was obtained, for $F = 50 \ \frac{mJ}{cm^2}$ and $t_T = 20 \ ps$, assuming thermal stress only (i.e. the hydrostatic pressure, p, is zero) and ignoring the first two terms in Eq. (5.4.43) as well as the generation of defects by photochemical bond breakage (first term in Eq. (5.4.44)). Results showed that the concentration of defects as a function of depth into the bulk increased substantially with a decrease in pulse duration from 150 ps to 10 ps. Thus, the effect of stress becomes more important for very short laser pulse, and it acts to enhance photophysical ablation.

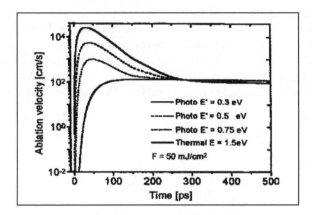

Fig. 5.4.7. Ablation velocity as a function of time for the photophysical and thermal models in the case of a single 500 fs laser pulse with a fluence of $\frac{50\text{mJ}}{\text{cm}^2}$. The activation energy in the ground state for all curves is E=1.5eV (reprinted from Luk'yanchuk et al. 1998 with permission from the SPIE)

Luk'yanchuk et al. (1998) compared the predictions of a photophysical and thermal models for ultra-short laser pulses (where the pulse duration is shorter than the relaxation time of excited species) in an effort to distinguish between them experimentally. The photophysical model was essentially that described in Eqs. (5.4.25), (5.4.27), (5.4.29), and (5.4.30) with the same initial and boundary conditions, and with k and c_p considered as functions of temperature. For the photothermal model, the ablation velocity and flux are given in Eqs. (5.4.27) and (5.4.31) with $N_s = 0$. The main features that showed a striking distinction between photophysical and pure thermal ablation were the ablation velocity per pulse as a function of time and the ablation depth per pulse vs relaxation time. Figure (5.4.7) shows that the ablation velocity for the thermal model increased monotonically with time up to a limit. This is explained by the increase in the surface temperature caused by relaxation of excited species followed by surface cooling resulting from heat diffusion and perhaps from the movement of the ablation front into the colder part of the bulk. The sharp increase in the ablation velocity to a maximum followed by a decrease to almost the same level as that of the thermal model can be explained as being caused by the formation of excited species followed by a decrease in their concentration over time. Figure (5.4.8) showed a substantial

increase in ablated depth with an increase in relaxation time for photophysical ablation in contrast to a slight decrease for the thermal model. This result
shows that the two models are substantially different in their predictions of
ablation depth for materials with long relaxation time and relatively high
activation energy for pure thermal ablation.

As is seen above, photophysical ablation can explain the relatively low
activation energy (about $1eV$) observed for polyimide (Kueper et al. 1993)
by considering, in addition to thermal ablation, a mechanism whereby ablation can occur as a result of desorption of excited species. Himmelbauer et al.
(1997) proposed that low activation energy can be explained also by a thermal model where mass loss of volatile species, produced by photodegradation
within the bulk, contributes to the ablation process. Letting N_l designate the
number density of volatile species thermally generated within the bulk, it is
seen that

$$\frac{\partial N_l}{\partial t} = v\frac{\partial N_l}{\partial z} + k_l(N_0 - N_l)\exp(-\frac{\Delta E}{k_B T}). \tag{5.4.49}$$

Also,

$$\frac{\partial I}{\partial z} = -\alpha I,$$

and

$$\frac{\partial T}{\partial t} = v\frac{\partial T}{\partial z} + D_\tau\frac{\partial^2 T}{\partial z^2} + \frac{I\alpha}{c_p\rho}. \tag{5.4.50}$$

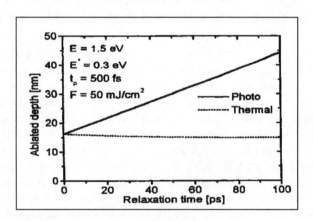

Fig. 5.4.8. Ablated depth as a function of relaxation time of excited species in the
case of photophysical and pure thermal ablation. E is activation energy in the
ground state and E* in the excited state (reprinted from Luk'yanchuk et al. 1998
with permission from the SPIE)

The ablation velocity is as before

$$v = v_A \exp(-\frac{\Delta E_A}{k_B T_s}), \tag{5.4.51}$$

and the boundary conditions are

$$\kappa \frac{\partial T}{\partial z} \big|_{z=0} = \rho v \Delta H = \rho v \frac{\Delta E}{M}, \tag{5.4.52}$$

$$N_l(z \to \infty) = 0, \ T(z \to \infty) = T_\infty,$$
$$I(z = 0) = I_0(t) \exp\left[-\alpha_g \int_0^t v(t')dt'.\right] \tag{5.4.53}$$

The initial conditions are

$$N_l(t = 0) = 0, \ T(t = 0) = T_\infty. \tag{5.4.54}$$

Here, attenuation of the laser intensity by the plume is considered as is seen from Eq. (5.4.53).

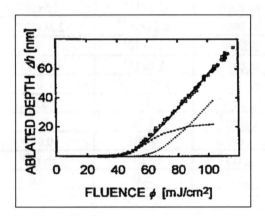

Fig. 5.4.9. Ablated depth for polyimide in vacuum caused by thermal ablation (dotted curve) and by elimination of volatile species (dashed curve). The solid curve is the sum of both curves as obtained from Eq. (5.4.55) in the text (reprinted form Himmelbauer et al. 1997 with permission from the SPIE)

The ablation depth is calculated from the total mass loss as

$$\Delta M = \rho \int_0^\infty v dt + m \int_0^\infty N_l(x, \infty)dx, \tag{5.4.55}$$

where m is the average mass loss per volatile (or gaseous) molecule. Here, mass loss is a combination of loss caused by thermal ablation and by the elimination of volatile species. A fit of the model Eqs. (5.4.49)-(5.4.55) to polyimide data ablated in vacuum by XeCl laser at 308 nm is shown in Fig. (5.4.9). The figure shows an excellent fit of the model to the data. The solid line in the figure is the sum of the dashed line (mass loss caused by volatile species) and the dotted line (mass loss caused by thermal ablation). Parameters used in fitting the model are given in Table (5.4.2). However, model fit to polyimide data ablated in air by an Ar^+ laser at 302 nm with pulse lengths of 140 and 2000 ns did not show as good a fit. The predicted ablated depth was somewhat larger than what was observed.

Table 5.4.2. Parameters used in the numerical calculations of the model (reprinted from Himmelbauer et al. 1997 with permission from the SPIE)

	XeCl λ=308 nm	Ar^+ λ=302±4.5 nm
ΔE_A, eV	2.75	2.75
$\Delta E_I^{\bullet}$, eV	1.47	1.47
v_A, cm/s	1.5×10^6	1.5×10^6
$k_I^{\bullet}$, cm/s	2.24×10^{10}	2.24×10^{10}
ΔH, J/g	3.3×10^3	0
α, cm^{-1}	8.6×10^4	1.26×10^5
α_z, cm^{-1}	1.035×10^5	0
c_p, J/g·K	1.1	1.1
ρ, g/cm^3	1.42	1.42
D_T, cm^2/s	0.001	0.001
N_{α}, cm^{-3}	6×10^{21}	6×10^{21}
T_{∞}, K	300	300

A model proposed by Schmidt et al. (1998) considers a system of chromophores with an excited and a ground state and combines in the ablation process photothermal decomposition and photodissociative bond breaking in the excited species. Incorporated in the model are laser attenuation by the plume, laser-induced modification of the absorption coefficient, and heat conduction. It is assumed that a species in the excited state can return to the ground state (i.e., undergo internal conversion (IC)) or can undergo direct decomposition (dd) caused by bond breakage. Internal conversion leads to species in the ground state that are highly vibrationally excited. This vibration can cause bond breakage or can relax with time and dissipate into heat. Chromophores that undergo bond breakage in the excited state return to the ground state as modified chromophores with a different absorption coefficient from the original chromophores. For calculation purposes, the polymer is treated as a system of infinitesimally thin layers each with chromophore concentration n_0 in the ground state and n_1 in the excited state. The change in n_1 with time is given by

$$\frac{dn_1}{dt} = \frac{n_0 \sigma_{eff} I}{h\nu} - k_{IC} \times n_1 - k_{dd} \times n_1, \qquad (5.4.56)$$

where k_{IC} and k_{dd} are the internal conversion and direct decomposition coefficients and σ_{eff} is the effective absorption cross-section in each layer defined as

$$\sigma_{eff} = \sigma_{or} + \frac{(\sigma_{\mathrm{mod}} - \sigma_{or})n_{\mathrm{mod}}}{n_{th}}, \qquad n_{\mathrm{mod}} \le n_{th}, \qquad (5.4.57)$$

$$\sigma_{eff} = \sigma_{\mathrm{mod}}, \quad n_{\mathrm{mod}} > n_{th}, \qquad (5.4.58)$$

where n_{mod} is the concentration of modified chromophores, n_{th} is the threshold density, σ_{mod} is the absorption cross-section of the modified chromophores, and σ_{or} of the original chromophores. The thin polymer layer is considered to be ablated when n_{mod} reaches or exceeds n_{th}. The concentration of modified chromophores can be expressed as

$$\frac{dn_{\mathrm{mod}}}{dt} = k_{dd}\frac{n_c - n_{\mathrm{mod}}}{n_c}n_1 + k_T\frac{n_c - n_{\mathrm{mod}}}{n_c}n_0, \qquad (5.4.59)$$

where n_c is the chromophore density. The first term in Eq. (5.4.59) is caused by bond breakage in the excited state as well as in the ground state after internal conversion and before relaxation. The second term is caused by thermal decomposition in the heated material. The thermal decomposition coefficient k_T is given by the Arrhenius equation. The attenuation of the laser intensity with depth is given by

$$\frac{dI}{dx} = -f n_0 \sigma_{eff} I. \qquad (5.4.60)$$

Here, $f\sigma_{eff}$ (where $f = \left(1 + \frac{vt}{r}\tan(\beta)\right)^{-2}$ and $r = $ beam radius) may be considered as the total absorptivity of the ablated material during its expansion (limited laterally to an angle β, measured as the angle between the lateral direction of movement and the normal to the surface).

The temperature in each polymer layer may be caused by internal conversion followed by vibrational relaxation and by thermal decomposition in the ground state. This gives rise to the equations

$$\frac{dT}{dt} = k_{IC} \times n_1 \frac{h\nu}{c_p}, \qquad (5.4.61)$$

and

$$\frac{dT}{dt} = \left(k_T\frac{n_c - n_{\mathrm{mod}}}{n_c}n_0\right)\frac{\Delta H}{c_p}, \qquad (5.4.62)$$

where ΔH is the reaction enthalpy. Also, the temperature change caused by heat conduction from a given layer to other layers in the x direction (depth) is given by the heat equation

$$\frac{dT}{dt} = \frac{\chi}{c_p}\frac{d^2T}{dx^2},$$ (5.4.63)

where χ is the heat conductivity. The heat capacity, c_p, and heat conductivity were expressed as linear functions of T.

Model equations were fitted to data ablated by excimer laser at 248, 308 nm for polyimide, and 248 nm for PMMA. Model parameters chosen and justifications are discussed in Schmidt et al. (1998). The model shows good agreement with experimental data for different pulse durations, Fig. (5.4.10), and different beam spot size, Fig. (5.4.11). It is seen from these figures that the ablation rate increased with an increase in pulse duration or with a decrease in spot size. An interesting finding concerning the plume effect was the fact that ignoring plume expansion by setting $\beta = 0$ eliminated the effect of spot size or pulse duration on the ablation rate. The model predicts relatively low surface temperatures during ablation (2000 to 3600 K for polyimide and 450 to 750 K for PMMA). These temperatures were in agreement with reported experimental observations in the literature.

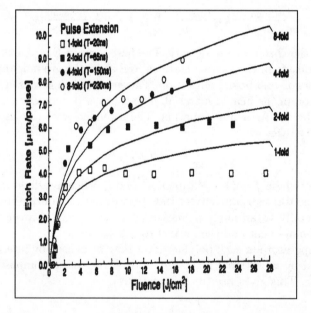

Fig. 5.4.10. Ablation rate as a function of fluence and pulse duration for 248 nm wavelength. (a) polyimide; (b) PMMA (reprinted from Schmidt et al. 1998 with permission from the American Institute of Physics)

A model considering plume formation as a first-order phase transition involving surface or bulk evaporation was developed by Afanasiev et al. (1997) and analytic expressions for etch depth per pulse were obtained. The governing equations for the plane expansion of the plume (under the assumption that the focal spot radius is much larger than the plume size) was expressed as (Zel'dovich and Raizer 1967)

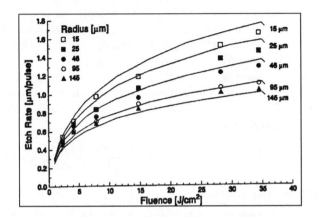

Fig. 5.4.11. Ablation rate for polyimide as a function of fluence for different beam spot sizes and for 248 nm wavelength (reprinted from Schmidt et al 1998 with permission of the American Institute of Physics)

$$\frac{\partial \rho}{\partial t} + \frac{\partial (\rho v)}{\partial x} = 0; \quad \frac{\partial (\rho v)}{\partial t} + \frac{\partial (p + \rho v^2)}{\partial x} = 0;$$

$$\frac{\partial [\rho(\varepsilon + v^2/2)]}{\partial t} + \frac{\partial [\rho v(\varepsilon + v^2/2 + p/\rho)]}{\partial x} - \frac{\partial q}{\partial x} = 0, \tag{5.4.64}$$

where ρ, v, p, and ε are the density, velocity, pressure, and specific energy of the plume, respectively. $q(x,t) = q_0 \exp[-\int_x^{x_{fr}} \alpha(x',t)dx']$ is the laser intensity, where $q_0 = q(\infty,t)$, and $\alpha(x,t)$ is the absorption coefficient of the plume. Here, $x_0 \leq x \leq x_{fr}$, where x_0 is the coordinate of the surface of the polymer film and x_{fr} is the coordinate of the front boundary of the plume. When the plume is considered to be transparent or nonabsorbing, then $\frac{\partial q}{\partial x} = 0$. The boundary conditions for the plume at $x = x_{fr}$ are given by

$$p(x_{fr}) = 0; \quad q(x_{fr}) = q_0. \tag{5.4.65}$$

The boundary conditions at $x = x_0$ are

$$\rho_1(v_1 + D) = \rho_0 D; \quad p_1 + \rho_0 D v_1 = p_0; \quad \rho_0 D(\varepsilon_1 + v_1^2/2) + p_1 v_1 = q_0, \tag{5.4.66}$$

where $D = -\frac{dx_0}{dt}$ is the velocity of the ablation surface, ρ_1, v_1, p_1, and ε_1 are the density, velocity, pressure, and specific energy of the plume phase at the boundary $x = x_0 + 0$; ρ_0 and p_0 are the density and pressure of the polymer at $x = x_0 - 0$. The ablation products are assumed to form a gas with an effective isoentropic exponent γ and specific energy $\varepsilon_1 = \frac{p_1}{(\gamma-1)\rho_1} + \frac{N_A U}{\mu}$, where N_A is the Avogadro number, μ is the molar mass of the ablation products, and U is the activation energy per particle. A solution to Eq. (5.4.64) with boundary conditions in Eqs. (5.4.65) and (5.4.66) is given in Afanasiev et al. (1997). Analytic expressions are derived for etch depth per pulse in the case of strong and weak absorbing polymers neglecting the plume absorption effect. Also, expressions for etch depth per pulse, considering radiation absorption by the plume, are given for plane and spherical expansions of the plume. Here, we consider expressions for etch depth per pulse for strong and weak absorbing polymers (ignoring the plume absorption effect) as compared with experimental results. For strong absorbing polymers such as polyimide, the depth per pulse is given by

$$d = D\tau = \frac{b^2 F}{(b^2 + \lambda_s^2)\rho_0 \Omega_0}, \tag{5.4.67}$$

where λ_s satisfies the relation $\lambda_s(b^2 + \lambda_s^2)\exp(-\frac{1}{\lambda_s^2}) = \frac{F}{F_0}$, $\Omega_0 = \frac{N_A U_0}{\mu}$, $b^2 = \frac{2(\gamma-1)}{\gamma(\gamma+1)}$, $\tau =$ pulse duration, and $F = q_0\tau$ is the pulse fluence. This indicates that the etch depth is linearly related to laser fluence. However, comparison to observed etched depth data for polyimide shows that linearity holds for fluences less than $\frac{3J}{cm^2}$. For higher fluence, the model overestimates the observed etch depth. This could be attributed to the plume screening effect which was ignored in the model. For weak absorbing polymers, the etch depth per pulse is expressed as

$$d = d_0\lambda_w; \quad \lambda_w = (\frac{F}{2F_0})^{1/3}[(a+1)^{1/3} - (a-1)^{1/3}],$$
$$a = \sqrt{1 + 0.15(F_0/F)^2}, \tag{5.4.68}$$

where $F_0 = \left[\frac{2(\gamma-1)}{\gamma+1}\right]^{\frac{1}{2}} \beta\rho_0\Omega^{\frac{3}{2}}\tau$, and $d_0 = \left[\frac{2(\gamma-1)}{\gamma+1}\right]^{\frac{1}{2}} \beta\Omega^{\frac{1}{2}}\tau$. Expression (5.4.68) shows that the etch depth per pulse is $d \sim F^{\frac{1}{3}}$. Comparisons with observed etch depth for PMMA at 248 nm wavelength and at pulse durations of 16 ns, 61 ns, and 149 ns revealed a good fit of the model for fluences less than about 5 or 6 $\frac{J}{cm^2}$. At higher fluences, the model overestimated the observed etch depth. This discrepancy may be caused again by the fact that the model ignored the plume screening effect.

Expressions for etch depth per pulse were derived in the case where absorption by the plume is taken into consideration. When absorption by the plume is assumed to be proportional to the plume mass, an expression for etch depth, under plane expansion of the plume, is given by

$$d = \frac{1}{\alpha}[1 - \exp(-\frac{\alpha F}{\rho_0 \Omega})], \tag{5.4.69}$$

where α is the absorption coefficient of the polymer. On the other hand, when absorption is according to Beer's law, the etch depth is expressed as

$$d = \frac{1}{\alpha_0} \ln(1 + \frac{\alpha_0 F}{\rho_0 \Omega}), \tag{5.4.70}$$

where $\alpha_0 = a\rho_0$ (a is a constant). Under the assumption of spherical plume expansion, the etch depth per pulse is given by

$$d = A(\gamma, \Omega_f) \frac{(F\tau^2/\rho_0)^{\frac{1}{3}}}{(ar_f)^{\frac{2}{3}}}, \tag{5.4.71}$$

where r_f is the radius of the focal spot, $\Omega_f = \frac{\pi r_f^2}{r_0^2}$ (r_0 is the target radius), and $A(\gamma, \Omega_f) = \frac{4}{\gamma+1}\left[(\gamma-1)\frac{\Omega_f}{\pi}\right]^{\frac{1}{3}}\left[1 + (\frac{\Omega_f}{\pi})^{\frac{1}{3}}\right]^{\frac{2}{3}}$. For more details the reader is referred to Afanasiev et al. (1997).

It is seen from the models discussed so far that the best model to use in an experimental situation depends on the polymer structure as well as on the laser parameters such as fluence, pulse duration, and laser wavelength. As such, it is necessary to test different models in order to determine the best one for any given situation. Singleton et al. (1990) evaluated the predictability of four laser ablation models discussed above, namely the models by Keys et al. (1985), Sauerbrey and Pettit (1989), Sutcliffe and Srinivasan (1986), and Lazare and Granier (1989). Predictions of etch depth versus fluence, for pulses of 5 ps, 7 ns, and 300 ns, were calculated from each of the four models and compared with experimental values (in the case of polyimide) for the same pulse lengths of an XeCl laser at 308 nm. It was found that the models of Keys et al. (1985), Eq. (5.3.13), and of Sauerbrey and Pettit (1989), Eq.(5.4.13), gave the best fit to the experimental data. The model by Sutcliffe and Srinivasan (1986) did not give accurate estimates of the threshold values for pulses of 5 ps and 300 ns, and the calculated etch depth from the model, Eq. (5.3.17), differed substantially from the experimental etch depth for 5 ps and 300 ns pulses. Also, this model was not in agreement with etch depth data at high fluences for PMMA at 16 ns pulse duration (Kueper and Stuke 1987) and for polytetrafluoroethylene (Teflon) at 300 fs pulse duration (Kueper and Stuke 1989). The model of Lazare and Granier (1989), given by the integral over pulse duration of Eq. (5.3.18) for constant parameters, I, k, and β, did not show good agreement with experimental data for threshold and etch depth per pulse at pulse durations of 5 ps and 300 ns. As pointed out by Singleton et al. (1990), it is of interest to note that the model Eq. (5.3.13) of Keys et al. (1985) has the same form as Eq. (5.4.13) of Sauerbrey and Pettit (1989). This is seen to be so if, at the onset of ablation, one replaces F in Eq. (5.3.13) by $exp(-k_1 E_{th})$ and rearranges to give

$$d = \frac{k_1}{k_2 N}(E - E_{th}) + \frac{1}{k_2 N} \ln\left(\frac{1 - e^{-k_1 E}}{1 - e^{-k_1 E_{th}}}\right), \tag{5.4.72}$$

where $E = I_0 t$. Here, $k_2 N = \alpha = \sigma_1 \rho_0$ and k_1 corresponds to σ_1 in Eq. (5.4.13). Also, the two models gave identical results.

Costela et al. (1995) used data from experiments on the ablation of PMMA and poly(2-hydroxyethyl methacrylate) or PHEMA, with 0, 1, and 20% of ethylene glycol dimethacrylate (EGDMA) added as a cross-linking monomer to test the performance of the models in Eqs. (5.3.18), (5.4.4), and (5.4.6). It was shown that the model in Eq. (5.3.18) gave the best fit to the etch depth per pulse versus excimer laser fluence for wavelengths of 193 nm, 222 nm, and 308 nm. The laser pulse width at half-maximum was approximately 7 ns for 222 nm, 13 ns for 193 nm, and 32 ns for 308 nm. However, the same model failed to fit the etch rate data at laser fluences higher than $\frac{0.6J}{cm^2}$ for the 193 nm wavelength, $\frac{2J}{cm^2}$ for 222 nm, and $\frac{8J}{cm^2}$ for 308 nm. In the high fluence range, the ablation mechanism seems to differ significantly from model assumptions. Estimates of k and β from model fit to the PMMA and PHEMA etch rate data indicated that both parameters depend on the laser wavelength and polymer structure. The parameter or ablation rate constant k increased with a decrease in wavelength. Also, β increased with a decrease in wavelength indicating a higher screening effect at shorter wavelength. These parameters give information about the polymer's ablation characteristics and how readily it can be ablated.

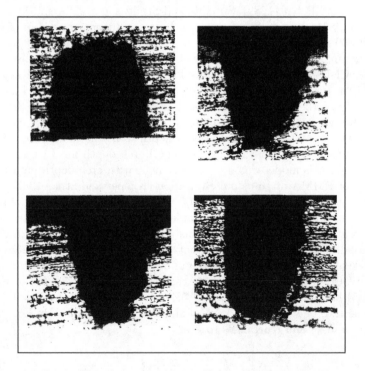

Fig. 5.4.12. Laser ablated microcavities in Neldin 2000 using a 351 nm excimer laser (reprinted from Nassar and Krishnan 1998 with permission of the SPIE)

As is seen from the models discussed in this chapter, modeling research on laser ablation has concentrated mainly on predicting the etch depth per pulse. Predicting etch depth per pulse is essential for developing models to be used in surface scanning operations to fabricate prototypes of a specified geometry. Figure (5.4.12) shows typical cases of ablated microcavities in polyimide (Neldin 2000, Furon, Inc, Bristol, RI) using a 351 nm pulsed laser, where the intension was to produce cavities with perpendicular walls. As is seen, the walls are not perpendicular, and no method exists at present other than trial and error that allows one to produce the intended geometry. In order to enhance the development of laser ablation as an accurate microfabrication tool, models must be developed to predict the surface scanning scheme of the laser beam based on the desired geometry of the microstructure to be fabricated. Nassar and Krishnan (1998) developed, as a first step in this direction, a simple model to predict the number of laser pulses at a pixel on the scanned surface in order to ablate a microstructure with a prespecified geometry.

Consider a surface divided into $n_1 \times n_2$ pixels. Let d_{ij} represent the etch depth per pulse at the $ijth$ pixel ($i = 1, 2, ..., n_1; \quad j = 1, 2, ..., n_2$). For ease of applications, it is advantageous to use an analytic function for d_{ij}. As is seen in this chapter, there are several analytic functions that one may use for d_{ij} depending on the experimental conditions such as polymer structure as well as on the laser parameters such as fluence, pulse length, inter-pulse duration, and laser wavelength. It is necessary then to select the function for a given experimental situation based on comparisons between predicted and experimental etch depth. For low fluence, functions to investigate are Eqs. (5.3.7) and (5.3.25). For a wide fluence range, Eqs. (5.4.69)-(5.4.71), Eqs. (5.2.1) (in combination with Eq.(5.2.4)), (5.4.4), (5.3.13), and (5.4.13) may be considered. Equations (5.4.13) and (5.3.13) (see Eq. (5.4.72)) seem to have wide applicability.

In general, one may express d_{ij} as a function of the incident laser fluence ($F_{i,j}$), $d_{ij} = f(F_{i,j})$. Because of the distribution of the laser beam on the (x, y) surface, the incident fluence received at pixel (x_i, y_j) from the laser beam focused at pixel (x_k, y_l) may be expressed as

$$F_{i,j} = F \times g_{x_k, y_l}(x_i, y_j) \Delta x \Delta y, \qquad (5.4.73)$$

where $g(x, y)$ is the probability density function, pdf, of the laser fluence and (x_i, y_j) represents the midpoint of the $ijth$ pixel. From Eq. (5.4.73), it is seen that

$$F \cong \sum_{i=1}^{n_1} \sum_{j=1}^{n_2} F_{i,j}. \qquad (5.4.74)$$

For a Gaussian distribution,

$$g_{x_k, y_l}(x_i, y_j) = \frac{1}{2\pi\sigma^2} \exp\left[-\left(\frac{(x_i - x_k)^2 + (y_j - y_l)^2}{2\sigma^2}\right)\right]. \qquad (5.4.75)$$

Considering all (x_k, y_l) pixels, it is seen that the etch depth $Z_{i,j}$ at pixel (x_i, y_j) is given by

$$Z_{i,j} = \sum_{k=1}^{n_1} \sum_{l=1}^{n_2} f(F_{i,j}) N_{k,l}. \qquad (5.4.76)$$

Here, $N_{k,l}$ is the number of pulses at pixel (x_k, y_l). In matrix notation, Eq. (5.4.76) is expressed as

$$\{H_{k,l}(i,j)\}\{N_{k,l}\} = \{Z_{i,j}\}, \qquad (5.4.77)$$

where $H_{k,l}(i,j) = f(F_{i,j})$. Here, $\{H_{k,l}(i,j)\}$ is a $n \times n$ matrix ($n = n_1 n_2$) and $\{N_{k,l}\}, \{Z_{i,j}\}$ are $n \times 1$ column vectors.

In Eq. (5.4.77), $\{H_{k,l}(i,j)\}$ can be determined. Also, the depth $Z_{i,j}$ is determined from knowledge of the prespecified geometry of the device to be etched. As such, Eq. (5.4.77), which is a linear system of n equations in n unknowns, can be solved numerically for the $\{N_{k,l}\}$ vector using techniques such as the Gauss-Seidel numerical method. The solution is approximate since $\{N_{k,l}\}$ may not be in integers. For an exact solution, one may have to account for a fraction of a pulse by adjusting pulse duration which may be difficult to obtain in practice. The vector $\{N_{k,l}\}$ determines the scanning pattern of the laser in order to etch with accuracy a microstructure of a desired geometry. Such models, with computer implementation to control the laser beam scanning pattern, are needed for rapid progress in the application of laser ablation to microfabrication.

5.5 High Energy Density Laser Models

In this chapter, our interest is mainly in relatively low energy density pulsed-laser ablation of polymers. However, a body of literature is concerned with laser cutting and drilling, mainly in metals, where lasers are employed at very high energy densities (in thousands of $\frac{W}{cm^2}$). In this section, we briefly highlight the main features of some of the models in this area.

A one-dimensional steady-state model that predicts material removal rate as a function of laser power has been developed by Chan and Mazumder (1987). The model is complex and involves heat conduction in the solid region, heat conduction in the liquid (molten material) region, fluid motion in the liquid region, and gas dynamics in the vapor. Also, a Knudsen layer is taken across the vaporization surface where discontinuities exist in temperature, density, and pressure. Solutions for material removal rates in dimensionless form were obtained. Figure (5.5.1) presents the predicted vaporization removal rate, liquid removal rate, and total (sum of both) removal rate in meter per sec for aluminum, superalloy, and titanium.: It is seen that the relative magnitude of vaporization and liquid removal rates varies with the laser power and the material being removed. For aluminum, the liquid expulsion rate is dominant at low power while the vaporization rate is dominant at high power. For superalloy, the vaporization rate is dominant over the whole power range, and for titanium the liquid expulsion rate is dominant.

Kar et al. (1992) considered a two-dimensional axisymmetric model with an a priori assumed cavity shape taking into account multiple reflections of the beam at the walls of the drilled cavity and assist gas-induced liquid metal flow during irradiation. It was found from the model that the effect of liquid metal flow on cavity depth was insignificant. On the other hand, the presence of multiple reflections caused the formation of deeper cavities. Also, cavity depth increased linearly with the gross laser intensity, P, ($P = \frac{pt}{\pi R^2 \tau}$, where p is the laser power, t is the irradiation time, R is the beam radius, and τ is the pulse duration) without reflection. However, this increase was somewhat non-linear with the presence of multiple reflections.

In a recent study, Ki et al. (2001) improved on the above models by implementing the level-set method mathematical technique (Sethian 1996) to track the evolution of the free moving liquid-vapor interface in order to determine the laser-created cavity profile over time. The model considers mass and energy balances in the liquid layer, fluid flow (using the Navier-Stokes equation), net mass loss caused by evaporation at the liquid-surface interface, multiple reflections inside the cavity, homogeneous boiling near the critical point, and tracking of the liquid-vapor interface. Figure (5.5.2) is a schematic presentation of the laser-material interaction. The equation of motion for level-set functions is given by

$$\frac{\partial \phi}{\partial t} + F_{lv} |\nabla \phi| + u_{adv} \nabla \phi = 0, \tag{5.5.1}$$

where ϕ represents the level-set functions, u_{adv} the fluid advection velocities, and F_{lv} the normal fluid velocities.

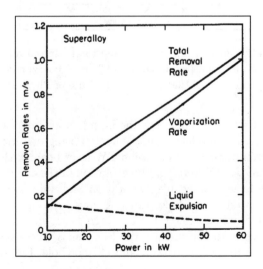

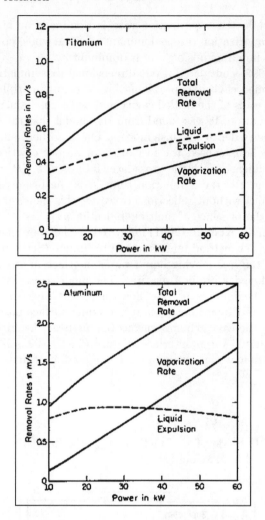

Fig. 5.5.1. Liquid expulsion, vaporization, and total removal rates for aluminum, superalloy, and titanium (reprinted from Chan and Mazumder 1987 with permission of the American Institute of Physics)

The zero level set (ϕ_0) represents the liquid-vapor interface that is of interest in the study. The model was implemented using material properties for a 1.5 mm thick iron plate and a Gaussian CW laser beam with an intensity of 10^7 and 10^8 W cm^{-2}. Of interest is the evolution of the cavity profile and the cumulative mass losses caused by evaporation and fluid flow. Figure (5.5.3) presents the evolution of the hole profile during exposure to the laser. It is seen that the profile is narrower for the high laser intensity. The effect of multiple reflections is to concentrate the laser energy near the center of the cavity leading to a deeper, narrower, and cleaner hole. Figure (5.5.4) shows the change in relative magnitudes of evaporative and fluid mass losses as a function of laser energy. At 10^7 W cm^{-2}, the mass loss rate caused by fluid flow is dominant. However, at 10^8 W cm^{-2}, evaporative mass loss becomes dominant with time. Also, it was seen from model results that the liquid layer thickness increased with an increase in the depth of the hole and decreased in thickness by about 60% with an increase in the laser energy density from 10^7 to 10^8 W cm^{-2}.

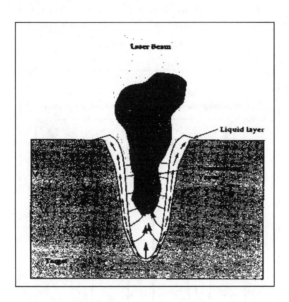

Fig. 5.5.2. Schematic diagram of the interaction process between a high energy density laser and material target (reprinted from Ki et al. 2001 with permission from IOP Publishing Limited)

In a similar modeling approach, using the level-set method, Kiel et al. (2002) investigated the effects of recoil pressure, multiple reflections, thermocapillary force, free surface evolution, laser power, and beam scanning in laser keyhole welding and compared theoretical predictions with experimental observations.

A model to describe groove depth and shape formed by evaporative cutting of a semi-infinite material target with a moving CW Gaussian beam laser has been considered by Modest and Abakians (1986). The authors assumed, as a first approximation to the problem, that the surface or beam moves at a constant velocity, the solid is opaque with no appreciable penetration of the laser into the solid, the solid is isotropic with temperature independent properties, change from solid to vapor occurs in one step, no plume absorption effect occurs, no multiple reflections of laser radiation occurs within the groove, and heat losses to the outside can be described by a constant heat transfer coefficient for both convective and radiative losses. Figure (5.5.5) is a schematic diagram of the three regions identified in the model. Here, the movement of the beam or the solid target is along the x-axis, and the coordinate system is fixed at the center of the beam. Region I is the flat part of the surface not yet exposed to the laser beam. It is bounded in the x, y directions by $x_{\min}$ and $y_{\max}$ ($x < x_{\min}$, $y > y_{\max}$), where $y_{\max}$ is the half-width of the final groove. Region II (evaporation zone) is close to the center of the Gaussian beam and lies between $x_{\min}(y)$ and $x_{\max}(y)$, ($x_{\min}(y) < x < x_{\max}(y)$). Region III is the fully established part of the groove where evaporation has taken place. Here, $x > x_{\max}(y)$.

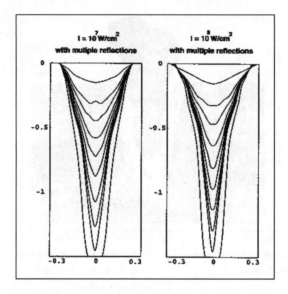

Fig. 5.5.3. Evolution of the hole profile during laser drilling (reprinted from Ki et al. 2001 with permission from IOP Publishing Limited)

Based on the above conditions, the heat equation in steady state can be expressed as

$$\rho c_p u \frac{\partial T}{\partial x} = k \nabla^2 T, \qquad (5.5.2)$$

where ρ, c_p, k, and u are the density, specific heat, thermal conductivity, and laser scanning speed, respectively. Equation (5.5.2) is subject to the boundary conditions $T \to T_\infty$ at $x \to + \infty$, $y \to + \infty$, $z \to \infty$. The boundary condition at the surface is given by

$$\alpha F_0(\hat{n}.\hat{k}) \exp\left(-\frac{x^2 + y^2}{R^2}\right) = h(T - T_\infty) - k(\hat{n}.\nabla T) - \rho h_{ig} u(\hat{i}.\hat{n}), \quad (5.5.3)$$

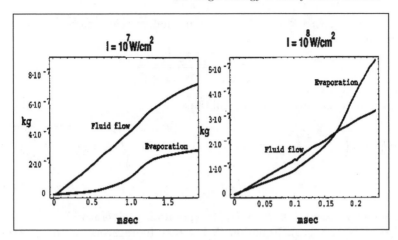

Fig. 5.5.4. Cumulative mass losses caused by evaporation and fluid flow as a function of time (reprinted from Ki et al. 2001 with permission from IOP Publishing Limited)

where α, h, h_{ig}, R, $\hat{\imath}$, $\hat{k}$, and $\hat{n}$ are the absorption coefficient, convective heat transfer coefficient, heat of sublimation, laser beam radius, unit vector in the x-direction, unit vector in the z-direction, and unit vector normal to the surface of the groove, respectively. It is seen that in region I the evaporation term is zero ($\hat{\imath}.\hat{n} = 0$) and $\hat{n} = \hat{k}$. Hence, Eq. (5.5.3) reduces to

$$\alpha F_0 \exp\left(-\frac{x^2 + y^2}{R^2}\right) = h(T - T_\infty) - k\frac{\partial T}{\partial z}; \quad x < x_{\min}(y). \tag{5.5.4}$$

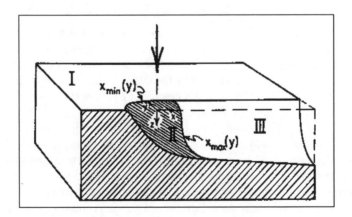

Fig. 5.5.5. Schematic diagram of the identified regions in evaporative laser cutting (reprinted from Modest and Abakians 1986 with permission from the ASME)

In region II, $(\hat{i}.\hat{n}) < 0$ and the unit normal to the surface $z = s(x, y)$ is seen to be

$$\hat{n} = \frac{(-\frac{\partial s}{\partial x} - \frac{\partial s}{\partial y} + 1)}{\sqrt{1 + (\frac{\partial s}{\partial x})^2 + (\frac{\partial s}{\partial y})^2}}.$$

Substituting $\hat{n}$ into Eq.(5.5.3), one obtains

$$\alpha F_0 \exp\left(-\frac{x^2 + y^2}{R^2}\right) = (h(T - T_\infty) - k\hat{n}.\nabla T)\sqrt{1 + (\frac{\partial s}{\partial x})^2 + (\frac{\partial s}{\partial y})^2}$$

$$+ \rho h_{ig} u \frac{\partial s}{\partial x}, \tag{5.5.5}$$

with boundary condition $T = T_{ev}$ (evaporation temperature), $z = s(x, y)$, $x_{\min}(y) < x < x_{\max}(y)$. Equation (5.5.5) may be regarded as the governing equation for the groove depth, $s(x, y)$. In region III, $\hat{i}.\hat{n} = 0$ and the unit normal to the surface $z = s_\infty(y)$ becomes

$$\hat{n} = \frac{(-\frac{\partial s_\infty}{\partial y} + 1)}{\sqrt{1 + (\frac{\partial s_\infty}{\partial y})^2}}.$$

Hence, Eq.(5.5.3) reduces to

$$\alpha F_0 \exp\left(-\frac{x^2 + y^2}{R^2}\right) = (h(T - T_\infty) - k\hat{n}.\nabla T)\sqrt{1 + (\frac{\partial s_\infty}{\partial y})^2}; \quad x > x_{\max}(y),$$

$$\tag{5.5.6}$$

where $z = s_\infty(y)$ is the depth of the fully established groove. The above set of equations has been transformed into non-dimensional forms by introducing the following non-dimensional variables and parameters:

$$\xi = \frac{x}{R}, \quad \eta = \frac{y}{R}, \quad \zeta = \frac{z}{R}; \quad S = \frac{s(x, y)}{R}, \quad \theta = \frac{(T - T_\infty)}{(T_{ev} - T_\infty)}, \quad N_e = \frac{\rho u h_{ig}}{\alpha F_0},$$

$N_k = \frac{k(T_{ev} - T_\infty)}{\alpha R F_0}$, $Bi = \frac{hR}{k}$, $U = \frac{\rho c_p u R}{k}$. The governing equations were solved numerically to determine the effects of laser and solid parameters on the groove depth and shape. Results showed that surface heat losses had no substantial effects on groove depth and shape. On the other hand, heat conduction and scan speed had significant effects. Figure (5.5.6) shows the effect of conduction loss, N_k, on the depth and shape of the groove. It is seen that the groove is shallow for large N_k or high heat loss caused by conduction and becomes deeper as N_k (or heat loss) decreases. As expected, the groove depth decreases in cross-section as η (or y) increases. The groove is similar to a Gaussian shape, but as pointed out by the authors, it may have a sharp apex at $\eta = 0$ and is therefore not in full agreement with a Gaussian shape. Results also showed that increasing the scanning speed decreased the maximum groove depth.

In a series of papers, effects of relaxing some of the model assumptions were studied. Abakians and Modest (1988) reported on the effect of a semi-transparent medium (in contrast to an opaque medium) on the shape and depth of the groove. The laser heat source decay in the z direction was assumed to be proportional to $\exp(-\beta_s z)$, where β_s is the extinction coefficient in the solid. It is clear that as β_s approaches infinity, the entire laser energy is deposited on the surface as in the case of an opaque material. Results showed, as expected, that as β_s decreased, the depth of the established groove decreased. A decrease in the extinction coefficient reduces the surface temperature which leads to a reduced groove depth. Further investigations by Bang and Modest (1991) revealed that the effect of multiple reflections of laser radiation within the groove (where reflection was assumed to be in a diffuse fashion, valid for rough surfaces) was to increase the depth of the groove and to cause a flatter profile at the apex with steeper side walls of the groove cross-section. The material removal rate increased from 20% for material with low reflectivity to 70% for high reflectivity. Increase in groove depth with reflectivity is in agreement with results by Ki et al. (2001) with regard to laser drilling.

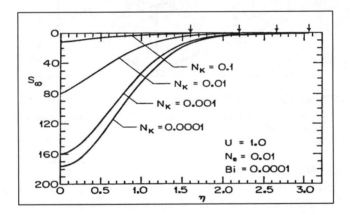

Fig. 5.5.6. Effect of heat conduction on the groove shape and depth (reprinted from Modest and Abakians 1986 with permission from the ASME)

Roy et al. (1993) extended the model by Modest and Abakians (1986) to the unsteady-state situation and allowed for variable properties of the solid. To solve the governing equations, they employed the finite-difference numerical technique with a coordinate transformation to map the complex groove geometry to a uniformly spaced rectangular coordinate region (boundary-fitted coordinate system). For shallow grooves, model predictions of groove shape and material removal rates were in fair agreement with experimental observations using silicon nitride (Si_3N_4). However, for deep grooves, the model underpredicts groove depth. This was attributed to the occurrence of multiple reflections of laser radiation within the groove which was not considered in the model. In a follow-up paper, Bang et al. (1993) extended the model to include multiple reflections of laser radiation within the groove. Diffuse (valid for rough surfaces) and specular (valid for smooth surfaces) reflections were considered in the model. In the case of deep grooves, there were better agreements between predicted and observed groove size and shape when diffuse reflections were incorporated in the model.

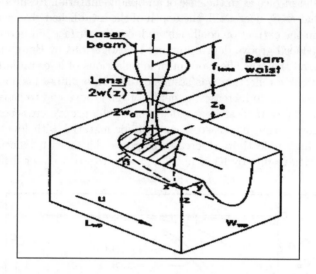

Fig. 5.5.7. Geometrical representation of laser and solid during laser beam scanning of the surface (reprinted from Modest 1996 with permission from Elsevier Science)

Modest (1996) used a similar 3-D heat conduction model to that in Roy et al. (1993) and investigated the development of groove shape formed by a laser in CW or in a pulsed mode with spatial and temporal beam distributions. The assumptions involved are basically those in Modest and Abakians (1986). The geometric representation of the interaction of the laser with the solid material is seen in Fig. (5.5.7). The 3-D heat conduction equation at the unsteady state is given by

$$\rho c_p \frac{\partial T}{\partial t} = \nabla(k\nabla T) \tag{5.5.7}$$

subject to the boundary conditions, in the case of a semi-infinite slab,

$$(T = T_\infty : x \to \pm\infty, \ y \to \pm\infty, \ z \to \infty), \tag{5.5.8}$$

$$\alpha F\hat{n} = -\hat{n}(k\nabla T) + v_n \rho h_{re} \ : \ z = s(x,y) \tag{5.5.9}$$

where h_{re} is the heat of removal or evaporation and v_n is the ablation velocity of the solid surface. The initial condition is

$$s(x,y,z,0) = 0, \ T(x,y,z,0) = T_\infty. \tag{5.5.10}$$

The boundary condition in Eq. (5.5.9) states that the laser energy absorbed at the surface of the solid is balanced by loss through conduction into the solid and by evaporation. It is clear that the ablation velocity v_n is zero in surface areas where no evaporation takes place. The laser fluence F for a moving laser with constant velocity u and with a Gaussian distribution, as modified from Kogelnik and Li (1956) when $u \neq 0$, is expressed as

$$F = \frac{2P}{\pi w^2(z)} \phi(t) \exp\left\{-\frac{2[(x-ut)^2 + y^2]}{w(z)^2}\right\} \left(\frac{(x-ut)\hat{\imath} + y\hat{\jmath}}{\sqrt{r_c^2(z) - x^2 - y^2}} + \hat{k}\right),$$
(5.5.11)

where P is the laser power, $w^2(z) = w_0^2 + \beta_\infty^2(z-z_0)^2$ is the beam radius when z is not at the focal plane z_0, w_0 is the $\frac{1}{e^2}$ beam radius at $z = z_0$, $\beta_\infty = \frac{\lambda}{\pi w_0}$, $r_c(z) = (z-z_0)\left[1 + \frac{w_0^2}{\beta_\infty^2(z-z_0)^2}\right]$, and $\phi(t)$ is the intensity distribution over pulse duration. For a CW laser, $\phi(t) = 1$. To solve Eq. (5.5.7) for the given boundary and initial conditions, v_n was expressed in terms of an Arrhenius type equation, namely

$$v_n = c_1 \exp(c_2(1 - \frac{T_{re}}{T})),$$
(5.5.12)

where T_{re} is the equilibrium ablation temperature, c_1 is a constant, and $c_2 = \frac{h_{re}}{RT_{re}}$, where R is the gas constant of the ablated material or vapor. Solution to Eq.(5.5.7) was obtained using the boundary-fitted coordinate system as in Roy et al. (1993). Results on temperature development over time for CW and pulsed laser as well as comparison of grooves generated with CW and pulsed laser were presented. Of interest was the finding that for pulsed laser of short duration (pulse length about$100ns$) energy loss through conduction was negligible, resulting in larger material removals and deeper grooves as compared to the CW laser. Also, after the end of the pulse the material cooled off rapidly and returned to ambient temperature before the start of the second pulse. The same model was investigated (Modest 1997) for a thin slab of finite size allowing for surface heat loss by convection and radiation. Under these conditions, the boundary conditions for Eq. (5.5.7) were modified accordingly to give

$$\frac{\partial T}{\partial y} = 0; \ y = 0, \ y = W,$$
(5.5.13)

$$\frac{\partial T}{\partial x} = 0; \ x = 0, \ x = L,$$
(5.5.14)

$$-k\frac{\partial T}{\partial z} = h(T - T_\infty) + \epsilon\sigma(T^4 - T_\infty^4); \ z = D,$$
(5.5.15)

$$\alpha F\hat{n} = -\hat{n}(k\nabla T) + h(T - T_\infty) + \epsilon\sigma(T^4 - T_\infty^4) + v_n\rho h_{re}; \ z = s(x, y, t).$$
(5.5.16)

Here W, L, and D are the width, length, and thickness of the slab, respectively. It was found that surface heat loss through convection and radiation had no significant effect on groove or temperature development.

6

Thin Films

6.1 Introduction

Thin films are often encountered in micro systems. They are important in semiconductor devices, in microelectromechanical devices such as sensors and actuators, and in masks. Heat transport through thin films is of vital importance in microtechnology applications. In some electronic applications, the reduction of the device size to microscale has the advantage of enhancing the switching speed of the device. On the other hand, size reduction increases the rate of heat generation which can lead to higher thermal load on the device. Heat transfer at the microscale is also important in the processing of materials with pulsed lasers. The high power of the laser on the surface of a metal film can result in a build-up of temperature on the film surface causing thermal damage. When a significant temperature rise occurs in a solid, deformation or highly elevated stress occurs caused by thermal expansion. Deformation and stress waves are major causes of thermal damage in laser processing of materials. Hence, studying the thermal behavior of thin films (in one or in multilayers) is useful for predicting the performance of a microelectronic device or for the manufacturing of microstructures. Heat transfer at the microscale cannot be properly described by the conventional heat transfer equation. In the classical theory of diffusion, the heat flux vector ($\mathbf{q}$) and the temperature gradient (∇T) across a volume of material are assumed to occur at the same instant of time. However, if the scale in one direction is at the microscale, the flux and temperature gradient in this direction will occur at different times. This time lag requires a new approach in the study of heat transfer at the microscale. In this chapter we discuss heat transfer models in single or multilayer thin films with time lags.

6.2 Heat Transport Models at the Microscale

Heat transport is governed by phonon-electron interactions in thin metallic films and by phonon collisions in insulators, dielectric films, and semiconductors. The energy state of a metal lattice is discretized into quanta called phonons. The energy state of a phonon is given by

$$E = hv, \tag{6.2.1}$$

where v is the vibrating frequency of the phonon at a certain temperature T and h is Planck's constant. Energy transport from one lattice to another occurs as a consequence of a series of phonon collisions over time. The mean free path d is defined as the average distance travelled by a phonon during two successive collisions. Also, the mean free time τ is the average of the times travelled by a phonon between two successive collisions. For a statistically meaningful estimate of free time and free distance, a large number of collisions must be considered. Heat transport models at the macroscopic scale assume a large physical domain where a large number of phonon collisions can occur. This large number of collisions translates into a sufficiently long time for heat transport to occur. The mean free time between two successive collisions, or the characteristic relaxation time, is of the order of picoseconds for metals. The mean free time for dielectric films and insulators is longer and is of the order of nanoseconds to picoseconds. Electrons have a mean free path of the order of 10^{-8} m at room temperature which is shorter than that for phonons.

Since physical dimensions at the microscale can be of the same order of magnitude as the mean free path and the heat transport response time is of the same order of magnitude as the mean free time, heat transfer models at the macroscale must be re-examined for their validity at the microscale. The concepts of temperature gradient and heat flux at the macroscale become questionable at the microscale. Temperature gradient in thin films may become discontinuous because of the lack of sufficient energy carriers (electrons and phonons) to affect a sufficiently large number of collisions between the two surfaces of the film. Also, if the temperature or heat flux response time in thin films is of the same order of magnitude as the mean free time or relaxation time, the effects of phonon interactions and scattering must be taken into consideration in heat transport at the microscale. Another consideration is the fact that a response time of the order of picoseconds ($10^{-12}s$) implies that the penetrating depth of heat carried by phonons travelling at the speed of sound (10^4 to $10^5 m\ s^{-1}$) is of the order of 10^{-8} to $10^{-7}m$. From the above discussion, it is clear that the microscale effects in space and time must be considered simultaneously in any heat transport model. A full discussion of this topic is presented in Tzou (1997).

When a metal is exposed to a short-pulse laser, photons from the laser beam first heat up the electron gas. The electron gas interacts next with the phonons causing a temperature rise in the metal lattice. A model describing heat transport in metal at the microscale based on phonon-electron interaction has been proposed by Qiu and Tien (1992, 1993) following earlier similar models by Kagnaov et al. (1957) and Anisimov et al. (1974). The model considers two coupled equations, one describing heating of the electron gas and the other heating of the metal lattice. As such, the governing equations are expressed as

$$C_e \frac{\partial T_e}{\partial t} = \frac{\partial}{\partial x}(\kappa \frac{\partial T_e}{\partial x}) - G(T_e - T_l), \tag{6.2.2a}$$

$$C_l \frac{\partial T_l}{\partial t} = G(T_e - T_l), \tag{6.2.2b}$$

where κ is the electron thermal conductivity , and C_e and C_l denote the heat capacity for electrons and metal lattice, respectively. Equation (6.2.2a) represents heating of the electrons by an external source of photons such as that from a laser. As a second step, the hot electrons heat up the metal lattice through phonon-electron interaction as described in Eq. (6.2.2b). Here, G is the phonon-electron coupling factor describing the exchange of energy between electrons and phonons. For T_e much higher than T_l, G is given by

$$G = \frac{\pi^2}{6} \frac{m_e n_e v_s^2}{\tau(T_e) T_e}, \tag{6.2.3}$$

where m_e is the electron mass, n_e is the electron number density, $\tau(T_e)$ is the electron relaxation time at temperature T_e, and v_s is the speed of sound given by

$$v_s = \frac{k}{2\pi h} (6\pi^2 n_a)^{-\frac{1}{3}} T_D. \tag{6.2.4}$$

Here, k is the Boltzmann constant, n_a is the atomic number density, and T_D is the Debye temperature. The electron thermal conductivity is expressed as

$$\kappa = \frac{\pi^2 n_e k^2 \tau_e T_e}{3m_e}. \tag{6.2.5}$$

Solving for m_e from Eq. (6.2.5) and substituting the result into Eq. (6.2.3) gives an alternate expression for G,

$$G = \frac{\pi^4 (n_e v_s k)^2}{18\kappa}. \tag{6.2.6}$$

Values of G for different metals are given in Qiu and Tien (1992). They range in value from $2.6 \times 10^{16} \frac{W}{m^3 K}$ for gold to $648 \times 10^{16} \frac{W}{m^3 K}$ for vanadium. A solution to Eqs. (6.2.2a,b) was obtained by Qiu and Tien (1992) for a gold film by employing the Crank-Nicholson finite difference scheme. For comparisons with experimental results, the normalized change in the transient temperature of the electron gas ($\frac{\Delta T_e}{(\Delta T_e)_{max}}$) was equated to the normalized change in reflectivity on the film surface ($\frac{\Delta R}{(\Delta R)_{max}}$) which was measured by the front-surface-pump and back-surface-probe technique (Brorson et al. 1987; Elsayed-Ali 1991; Qui et al. 1994).

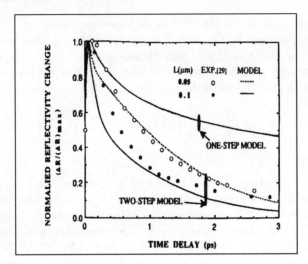

Fig. 6.2.1. Temperature change at the front surface of two gold films (.05 and 0.1 μm in thickness) exposed to a pulsed laser irradiation with pulse width of 96 fs, and fluence of $1 \frac{\text{mJ}}{\text{cm}^2}$. Reprinted from Qiu and Tien 1992 with permission from Eslevier Science. Experimental data from Brorson et al. 1987 with permission from the American Physical Society)

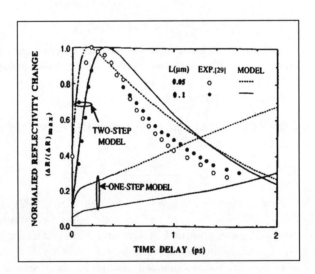

Fig. 6.2.2. Temperature change at the back surface of two gold films (.05 and 0.1 μm in thickness) exposed to a pulsed laser irradiation with pulse width of 96 fs, and fluence of $1 \frac{\text{mJ}}{\text{cm}^2}$. Reprinted from Qiu and Tien 1992 with permission from Elsevier Science. Experimental data from Brorson et al. 1987 with permission from the American Physical Society)

Figures (6.2.1) and (6.2.2) present comparisons between the normalized temperature change, from the standard heat diffusion model at the macroscale and from the phonon-electron interaction model of Eqs. (6.2.2a,b) and the experimental values (measured as the normalized change in reflectivity) at the front and back surfaces of gold films. The films were irradiated by a 96 femtosecond $(10^{-15}s)$ pulsed laser at a laser fluence of $1\frac{mJ}{cm^2}$. The time delay on the x-axis is equivalent to the real transient response time. It is seen from Fig. (6.2.1) that the transient temperature at the front surface of the film agrees quite well with model predictions from Eqs. (6.2.2a,b). As expected, the surface temperature decreased with time, and the temperature level was higher for the thinner film. It is interesting to see that the classical heat diffusion model did not agree with the observed data, an indication of the inadequacy of macroscale theory in explaining heat transport at the microscale. Figure (6.2.2) shows the transient temperature change at the rear surface of the film. As in the case of the front film surface, agreement with experimental data is seen to be far better for the microscale phonon-electron model than for the classical diffusion model. The classical heat diffusion model at the macroscale completely fails to describe heat transport at the microscale.

The two coupled Eqs. (6.2.2a,b) could be combined to give a single equation for the lattice temperature and another for the electron temperature (Tzou 1995a, 1995b). Solving for T_e from Eq. (6.2.2b), substituting the result into Eq. (6.2.2a), and replacing $G(T_e - T_l)$ by its value from Eq. (6.2.2b) gives a new type of heat conduction equation at the microscale

$$\nabla^2 T_l + \left(\frac{C_l}{G}\right)\frac{\partial}{\partial t}\nabla^2 T_l = \left(\frac{C_l + C_e}{\kappa}\right)\frac{\partial T_l}{\partial t} + \left(\frac{C_l C_e}{\kappa G}\right)\frac{\partial^2 T_l}{\partial t^2}. \qquad (6.2.7)$$

The interesting thing about this equation is having a thermal wave term $(\frac{\partial^2 T_l}{\partial t^2})$ and a mixed derivative term $(\frac{\partial}{\partial t}\nabla^2 T_l)$ reflecting the combined effects of microscale phonon-electron interaction and macroscale diffusion. A single equation describing the electron temperature at the microscale can be obtained from Eqs. (6.2.2a,b) in a similar manner. Solving for T_l from Eq. (6.2.2a), substituting the result into Eq. (6.2.2b), and replacing $G(T_e - T_l)$ by its value from Eq. (6.2.2b) gives

$$\nabla^2 T_e + \left(\frac{C_l}{G}\right)\frac{\partial}{\partial t}\nabla^2 T_e = \left(\frac{C_l + C_e}{\kappa}\right)\frac{\partial T_e}{\partial t} + \left(\frac{C_l C_e}{\kappa G}\right)\frac{\partial^2 T_e}{\partial t^2}. \qquad (6.2.8)$$

It is of interest to observe that Eqs. (6.2.8) and (6.2.7) have the same form.

Based on the Boltzmann equation for interacting phonons, Guyer and Krumhansl (1966) derived the governing equations relating temperature and flux in a system where heat transport is mainly by phonon collision/scattering. These equations are given by

$$C_p\frac{\partial T}{\partial t} + \nabla q = 0 \qquad (6.2.9a)$$

$$\frac{\partial \mathbf{q}}{\partial t} + \frac{c^2 C_p}{3} \nabla T + \frac{1}{\tau_R} \mathbf{q} = \frac{\tau_N \, c^2}{5} \left[\nabla^2 \mathbf{q} + 2\nabla(\nabla \mathbf{q}) \right], \qquad (6.2.9b)$$

where τ_N and τ_R are the relaxation times in which momentum in the phonon system is conserved or lost, respectively; c is the average speed of phonons (or the speed of sound); T is film temperature; and $\mathbf{q}$ is the flux vector. Taking the derivative (∇) on both sides of Eq. (6.2.9b) and replacing $\nabla \mathbf{q}$ by $-C_p \frac{\partial T}{\partial t}$ from Eq. (6.2.9a) gives the single heat equation at the microscale (Joseph and Preziosi 1989; Tzou 1995a)

$$\nabla^2 T + \frac{9\tau_N}{5} \frac{\partial}{\partial t}(\nabla^2 T) = \frac{3}{\tau_R \, c^2} \frac{\partial T}{\partial t} + \frac{3}{c^2} \frac{\partial^2 T}{\partial t^2}. \qquad (6.2.10)$$

It is interesting to note that Eq. (6.2.10) based on phonon collisions has the same form as Eqs. (6.2.8) and (6.2.7) based on phonon-electron interactions.

Majumder (1993) proposed a phonon radiative transfer model in which the phonon intensity (I_ω) at a vibration frequency ω is given by the integro-differential equation

$$\frac{1}{v} \frac{\partial I_\omega}{\partial t} + \mu \frac{\partial I_\omega}{\partial x} = \frac{\frac{1}{2} \int_{-1}^{1} I_\omega \, d\mu - I_\omega}{\tau v}, \qquad (6.2.11)$$

where τ is the mean free time or relaxation time in phonon scattering, $v = v(\theta, \phi)$ is the phonon velocity vector defined by the azimuthal angles θ (angle between the x-axis and the direction of phonon propagation) and ϕ in a spherical coordinate system, and $\mu = \cos\theta$. Derivation of Eq. (6.2.11) is discussed in Majumder (1993) and in Tzou (1997). It is seen from this equation that the phonon intensity travels as a wave into the solid film with a one-dimensional velocity, $v_x = v \cos\theta$. By solving Eq. (6.2.11) for the phonon intensity I_w, the temperature distribution of the film can be calculated from the Bose-Einstein equation at equilibrium relating I_ω to the temperature $T(x)$ in the x-direction (film thickness),

$$I_\omega^0(T) = \frac{1}{2} \int_{-1}^{1} I_\omega \, d\mu = \sum_p v_p \frac{h\omega D(\omega)}{\exp\left[\frac{h\omega}{\kappa T(x)}\right] - 1}, \qquad (6.2.12)$$

where the summation is over the three phonon polarizations (p), κ is the Boltzmann constant, h is the Planck constant, and $D(\omega)$ is the density of states per unit volume in the frequency domain of the vibrating lattice. Joshi and Majumder (1993) solved Eq. (6.2.11) numerically for I_ω and obtained the temperature distribution $T(x)$ from Eq. (6.2.12). They compared the temperature distribution $T(x)$ (at the microscale in both space and time) to those obtained from the diffusion equation (macroscopic in time and space)

$$\frac{\partial T}{\partial t} = \alpha \nabla^2 T, \qquad (6.2.13)$$

and from the thermal wave equation (macroscopic in space and microscopic in time)

$$\tau \frac{\partial^2 T}{\partial t^2} + \frac{\partial T}{\partial t} = \alpha \nabla^2 T. \tag{6.2.14}$$

It is seen that when the phonon relaxation time $\tau = 0$ (phonon velocity $v \to \infty$), the thermal wave equation reduces to the heat diffusion equation. The boundary conditions used for solving the diffusion and wave equations are

$$T = T_1 \text{ at } x = 0, \quad \text{and} \quad T = T_0 \quad \text{at } x = L; \ T_1 > T_0, \tag{6.2.15}$$

and those for solving Eq. (6.2.11) are

$$I_\omega = I_\omega^0(T_1) \text{ at } x = 0, \ \mu > 0 \quad \text{and} \quad I_\omega = I_\omega^0(T_0) \quad \text{at } x = L, \ \mu < 0. \tag{6.2.16}$$

Here, $I_\omega^0(T_1)$ and $I_\omega^0(T_0)$ are the intensities at equilibrium corresponding to temperatures T_1 and T_0. Figure (6.2.3) presents the dimensionless temperature profiles (defined as $\theta = \frac{T-T_0}{T_1-T_0}$) as a function of the dimensionless space (defined as $\frac{x}{L}$). The dimensionless time was defined as $\tau = \frac{t}{(L/\nu)}$, where ν is the average speed of sound. The diamond film thickness is $L = 0.1 \ \mu m$, and the times are $\tau = 0.1, 1.0$, and steady state for (a), (b), and (c), respectively. It is seen that the three models differ significantly in their predictions of the transient temperature distribution. The thermal wave equation (hyperbolic) predicts sharp wavefronts not seen for the other models. Also, the level of the temperature profile is lower for the phonon model (EPRT) than for the diffusion (Fourier) and thermal wave models. At steady state, the diffusion and thermal models converge to the same solution which is different from that of the phonon model. The authors showed that for an acoustically thick film (film thickness is much greater than the mean phonon free path, $v\tau$), $L = 10\mu m$, all three models converged to the same solution at steady state, showing a linear decrease in temperature as a function of x (depth).

It is clear from the above models that the mean free path ($l = \tau v$) in electron and/or phonon collisions is a significant factor in heat transport at the microscale. According to Flik et al. (1991), micro-scale effects in the transverse direction of a film become important when the film thickness is less than about 7 times the mean free path. On the other hand, in the longitudinal direction, microscopic effects become important when the film thickness is less than about 4.5 times the mean free path. In general, the characteristic dimension for which micro-scale effects in heat transport become important depends on the thermal loading condition, thermal properties of the thin film, and geometric configuration of the microstructure (Flik et al. 1991; Tien and Chen 1994).

A simplification to the two-step model in Eqs. (6.2.2 a,b) was proposed by Al-Nimir and Arpaci (2000) for the case of thin films exposed to picosecond duration laser pulse. Their approach involves decoupling the electron and lattice equations by eliminating the electron thermal diffusion term, $\frac{\partial}{\partial x}(\kappa \frac{\partial T_e}{\partial x})$, as being insignificant. This makes the reduced partial differential equations easier to solve. It was argued that electron diffusion can be ignored for systems with high electron-phonon coupling where the nonequilibrium time period is too short for electron thermal diffusion to take place. As a rule, it was suggested that diffusion becomes negligible when the ratio of electron-phonon coupling to the electron thermal conductivity is much larger than one, $\frac{GL^2}{\kappa_e} >>$ 1, which is true for metals such as lead, silver, copper, and gold. Under this model simplification, the thermal behavior occurs in two stages. The first stage, of short duration, involves absorption of the laser energy by the electrons and transmission of this energy to the metal lattice. The decoupled equations (with a heat source) describing this process are given by

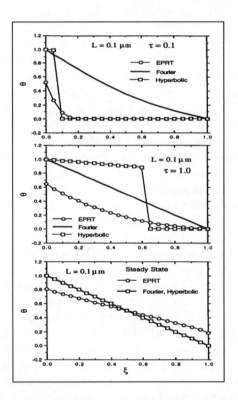

Fig. 6.2.3. Dimensionless temperature profiles (θ) as a function of dimensionless depth ($\xi = x/L$) predicted from the heat diffusion model (Fourier), the thermal wave model (Hyperbolic), and the phonon radiation model (EPRT) for different dimensionless times, $\tau = t/(L/\nu)$. The film thickness is 0.1 μm (reprinted from Joshi and Majumdar 1993 with permission from the American Institute of Physics)

$$C_e \frac{\partial T_e}{\partial t} = -G(T_e - T_l) + Q, \tag{6.2.17a}$$

$$C_l \frac{\partial T_l}{\partial t} = G(T_e - T_l). \tag{6.2.17b}$$

Thermal behavior in the film during the second stage is governed by the energy equation

$$\rho c \frac{\partial T}{\partial t} = -\frac{\partial q}{\partial x}, \tag{6.2.18a}$$

$$\bar{\tau} \frac{\partial q}{\partial t} + \kappa \frac{\partial q}{\partial x} + q = 0, \tag{6.2.18b}$$

where $\bar{\tau}$ is the relaxation time and $T = T_e = T_l$. Combining Eqs. (6.2.17 a,b) and substituting for T_l in the resulting equation, its value from Eq. (6.2.17a) gives

$$\frac{\partial}{\partial t} \left(\frac{C_e}{G} \frac{\partial T_e}{\partial t} \right) + \left(1 + \frac{C_e}{C_l} \right) \frac{\partial T_e}{\partial t} = \frac{\partial}{\partial t} \left(\frac{Q}{G} \right) + \frac{Q}{C_l}. \tag{6.2.19}$$

Also, combining Eqs. (6.2.18a,b) yields

$$\frac{\bar{\tau}}{\alpha} \frac{\partial^2 T}{\partial t^2} + \frac{1}{\alpha} \frac{\partial T}{\partial t} = \frac{\partial^2 T}{\partial x^2}. \tag{6.2.20}$$

The thermal behavior of a thin film in one-dimension is now governed by the solution to Eqs. (6.2.19) and (6.2.20).

6.3 Phase-Lag Models

Equations (6.2.7), (6.2.8), and (6.2.10) are universal equations describing heat transport at the microscale under various conditions or environments. They can be derived by an alternate approach utilizing the phase-lag concept (Tzou 1995a,b,c; Tzou 1997). This concept allows for the heat flux and the temperature gradient to occur at different instants of time in microscale heat transfer. This can be represented as

$$\mathbf{q}(r, t + \tau_q) = -\kappa \nabla T(r, t + \tau_T), \tag{6.3.1}$$

where r is the position of the flux vector $\mathbf{q}$ in the medium, τ_q and τ_T are the delay times for the heat flux and the temperature gradient, respectively. In the heat diffusion model at the macroscale, $\tau_q = \tau_T = 0$, which implies an infinite speed of heat propagation (zero delay time between heat input at one location and its detection at other locations). On the other hand, in the thermal wave theory of heat conduction, $\tau_T = 0$. A first-order expansion of Eq. (6.3.1) with respect to t (when $\tau_T = 0$), gives the classical thermal wave model

$$\mathbf{q}(r, t) + \tau_q \frac{\partial \mathbf{q}(r, t)}{\partial t} \cong -\kappa \nabla T(r, t) \tag{6.3.2}$$

as originally proposed by Cattaneo (1958) and Vernotte (1958). This wave equation relaxes the assumption of infinite speed of heat propagation. The heat flux delay time (or phonon mean free time) relates to the wave speed, C^2, by the expression (Chester 1963)

$$\tau_q = \frac{\alpha}{C^2}, \tag{6.3.3}$$

where α is thermal diffusivity. It is clear that when the wave speed is infinite, the wave equation reduces to the classical heat diffusion equation. In order to obtain a solution for $\mathbf{q}$ and T, Eq.(6.3.2) must be combined with the energy equation

$$-\nabla \mathbf{q}(r,t) + Q(r,t) = C_p \frac{\partial T}{\partial t}(r,t), \tag{6.3.4}$$

where Q is the heat source and C_p is the volumetric heat capacity. Note that Eq. (6.3.4) assumes instantaneous response between the temperature gradient and heat transport (instantaneous heat flow). Also, it is readily seen that taking the divergence of Eq. (6.3.2) and substituting the result into Eq. (6.3.4) gives the classical thermal wave equation with a source term

$$\frac{1}{\alpha}\frac{\partial T}{\partial t} = \nabla^2 T - \frac{\tau_q}{\alpha}\frac{\partial^2 T}{\partial^2 t} + \frac{1}{\kappa}\left(Q + \tau_q \frac{\partial Q}{\partial t}\right), \tag{6.3.5}$$

where $\alpha = \frac{\kappa}{C_p}$. With reference to the dual lag model in Eq. (6.3.1), the case $\tau_q < \tau_T$ implies that the flux (or heat flow) causes a temperature gradient to occur across the medium. On the other hand, when $\tau_q > \tau_T$, the temperature gradient (which occurs first) induces the flux or heat flow. A first-order Taylor series approximation to Eq. (6.3.1) leads to the expression

$$\mathbf{q}(r,t) + \tau_q \frac{\partial q}{\partial t}(r,t) \cong -\kappa\left\{\nabla T(r,t) + \tau_T \frac{\partial}{\partial t}[\nabla T(r,,t)\right\}. \tag{6.3.6}$$

As discussed by Tzou (1997), the lags or delayed responses in Eq. (6.3.1) can be caused by the following factors: (1) A lag is caused by the time it takes for heat to be transferred from the heated electrons to the phonons, which is usually of the order of picoseconds. (2) For dielectric films, insulators, and semiconductors, the dominant mechanism of heat transport is phonon collisions and scattering. In this case the lag is caused by the phonon mean free time. (3) Heat transport at low temperatures can be delayed because of a reduced collision rate among molecules. (4) Porous media can cause a delay in response caused by the time it takes for heat flow to circulate around the air pockets in the medium. As such, it is seen that the lags may belong to the intrinsic or to the structural properties of the material.

In order to obtain a unified temperature equation, one may combine Eqs. (6.3.6) and (6.3.4). Taking the derivative (∇) of Eq. (6.3.6) and substituting the solution for $\nabla \mathbf{q}(r,t)$ from Eq. (6.3.4) into Eq. (6.3.6) leads to the expression

$$\frac{1}{\alpha}\frac{\partial T}{\partial t} = \nabla^2 T + \tau_T \frac{\partial}{\partial t}(\nabla^2 T) + \frac{1}{\kappa}\left(Q + \tau_q \frac{\partial Q}{\partial t}\right) - \frac{\tau_q}{\alpha}\frac{\partial^2 T}{\partial t^2}. \qquad (6.3.7)$$

It is readily seen that Eq. (6.3.7) reduces to the diffusion equation when $\tau_q = \tau_T = 0$ and to the thermal wave Eq. (6.3.5) when $\tau_T = 0$. Furthermore, Eq. (6.3.7) has the same form as Eqs. (6.2.7), (6.2.8), and (6.2.10). Comparing Eq. (6.3.7) without the heat source term and the phonon scattering Eq. (6.2.10), it is seen that

$$\alpha = \frac{\tau_R c^2}{3}, \quad \tau_T = \frac{9\tau_N}{5}, \quad \text{and} \quad \tau_q = \tau_R. \qquad (6.3.8)$$

Also, when Eq. (6.3.7) is compared to Eq. (6.2.7), one obtains the relations

$$\alpha = \frac{\kappa}{C_e + C_l}, \quad \tau_T = \frac{C_l}{G}, \quad \text{and} \quad \tau_q = \frac{1}{G}\left[\frac{1}{C_e} + \frac{1}{C_l}\right]^{-1}. \qquad (6.3.9)$$

Hence, the dual-phase-lag model is a universal model that encompasses the microscale phonon-electron interaction model and the phonon scattering model as special cases.

Equations (6.3.4) and (6.3.6) may be combined by eliminating the temperature to give the following equation for flux:

$$\frac{1}{\alpha}\frac{\partial \mathbf{q}}{\partial t} = \nabla^2 \mathbf{q} + \tau_T \frac{\partial}{\partial t}(\nabla^2 \mathbf{q}) - \left(\nabla Q + \tau_T \frac{\partial}{\partial t}(\nabla Q)\right) - \frac{\tau_q}{\alpha}\frac{\partial^2 \mathbf{q}}{\partial^2 t}. \qquad (6.3.10)$$

Equation (6.3.10) is similar to Eq. (6.3.7) with $\mathbf{q}$ replacing T and is easier to use for problems involving flux boundary conditions. In the case of boundary conditions involving both flux and temperature, it is convenient to solve both Eqs. (6.3.4) and (6.3.6) simultaneously.

In deriving Eq. (6.2.7) (which is equivalent to the dual-phase lag equation in (6.3.7) without the heat source) from the two-step Eqs. (6.2.2a,b), heat conduction in the metal lattice was ignored. This was justified when one considers thin film exposure to short-pulse laser. Here, the time and space are small, and the model seems to be in excellent agreement with experimental data. When transient time and film thickness increase, conduction in the metal lattice may become important. Tzou (1997) derived a dual-phase lag equation accounting for conduction in the metal lattice. For temperature-independent heat conduction, Eqs. (6.2.2a,b) become

$$C_e \frac{\partial T_e}{\partial t} = \frac{\partial}{\partial x}\left(\kappa \frac{\partial T_e}{\partial x}\right) - G(T_e - T_l), \qquad (6.3.11a)$$

$$C_l \frac{\partial T_l}{\partial t} = k_l \nabla^2 T_l + G(T_e - T_l). \qquad (6.3.11b)$$

Solving for T_e from Eq. (6.3.11b), substituting the result into Eq. (6.3.11a), and replacing $G(T_e - T_l)$ by its value from Eq. (6.3.11b), one obtains the expression

$$-\left[\frac{k_e k_l}{G(k_e+k_l)}\right]\nabla^4 T_l + \nabla^2 T_l + \left(\frac{C_l k_e + C_e k_l}{G(k_e+k_l)}\right)\frac{\partial}{\partial t}\nabla^2 T_l$$

$$=\left(\frac{C_l+C_e}{k_e+k_l}\right)\frac{\partial T_l}{\partial t} + \left(\frac{C_l C_e}{G(k_e+k_l)}\right)\frac{\partial^2 T_l}{\partial t^2}. \tag{6.3.12}$$

In this equation, the highest order differential ∇^4 influences the characteristic of the solution which makes it different from the case where metal lattice conduction is ignored. Equation (6.3.12) can be expressed in a dual-phase lag (τ_q, τ_T) form to give

$$-(\alpha_p\tau_T)\nabla^4 T_l + \nabla^2 T_l + \tau_T\frac{\partial}{\partial t}\nabla^2 T_l = \frac{1}{\alpha}\frac{\partial T_l}{\partial t} + \frac{\tau_q}{\alpha}\frac{\partial^2 T_l}{\partial t^2}. \tag{6.3.13}$$

Here, $\tau_T = \frac{C_l k_e + C_e k_l}{G(k_e+k_l)}$, $\frac{1}{\alpha} = \frac{C_l+C_e}{k_e+k_l}$, $\frac{\tau_q}{\alpha} = \frac{C_l C_e}{G(k_e+k_l)}$ with $\tau_q = \frac{1}{G}\left[\frac{1}{C_l}+\frac{1}{C_e}\right]^{-1}$, and $\frac{1}{\alpha_p} = \frac{1}{\alpha_l} + \frac{1}{\alpha_e}$ with $\alpha_{e(l)} = \frac{k_{e(l)}}{C_{e(l)}}$.

6.4 Solution Methods

Methods of solution regarding phase lag-models in one dimension have been discussed by Tzou (1997). They include the method of Laplace Transform (with the Bromwich contour integrations and the Riemann-sum approximation utilized in the Laplace inversion) and the method of Fourier transform. Lin et al. (1997) obtained an exact solution to Eq. (6.3.7) in the one-dimensional case (where the heat source was taken to be a step function or a sudden rise in surface temperature to a constant value T_w), using the separation of variable technique. Antaki (1998) obtained a solution to the one-dimensional two-phase-lag equation using the Fourier sine transform to eliminate the x derivative followed by the Laplace transform to eliminate the time derivative. This method gave a double-transform solution of temperature. The solution was then obtained by taking the inverse Laplace transform followed by the inverse Fourier sine transform. The same one-dimensional two-phase-lag equation was solved by Tang and Araki (1999) using the Green's function method and the finite integral transform technique. For many applications, especially those involving two or three dimensions and/or non-constant thermophysical properties, analytical methods become intractable, and one must resort to numerical solutions.

Smith et al. (1999) used the separation of variable technique to solve the two-step model in Eqs. (6.2.2a,b) assuming a pulse heat source or a delta function and compared solution results to those obtained numerically using the Crank-Nicholson method. The delta function approximation of the heat source was based on the consideration that the short pulse duration is negligible on the time scale if it is considerably less than the thermalization time (time it takes for the temperature difference between electron and lattice to decrease by a factor $\frac{1}{e}$). As such, the heat source was expressed as

$Q = (1 - R)\frac{J}{d}\exp(-\frac{x}{d})\delta(t)$, where $\delta(t)$ is the delta function, R is reflectivity, and d is radiation penetration depth. The numerical solution was for constant thermophysical properties and for temperature-dependent properties. The latter included the electron heat capacity which was expressed as $C_e(T) = C_e T_e$ and the heat conductivity expressed as $\kappa(T_e, T_l) = \kappa(T_0)\frac{T_e}{T_l}$. Figure (6.4.1) presents conditions under which constant thermophysical properties can be assumed. These conditions are subject to 10% error. The dimensionless parameters in the Figure are $S = \frac{(1-R)J}{C_e T_0 d}$, $N = \frac{t_p}{\tau}$ (τ being the thermalization time), and $D = \frac{\kappa}{Gd^2}$, where G is the electron-phonon coupling factor. It is seen that constant thermophysical properties can be assumed for moderate to low fluence values as expressed by S for different N and D values.

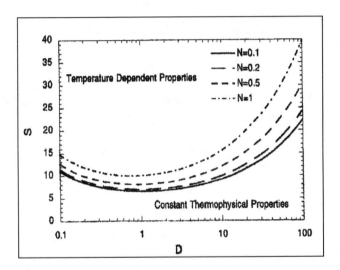

Fig. 6.4.1. Regions in which one may assume constant thermophysical properties. Here, $D=\frac{\kappa}{Gd^2}$, $N=\frac{t_p}{\tau_e}$, and $S=\frac{(1-R)J}{C_e T_0 d}$ (reprinted from Smith et al. 1999 with permission from Taylor & Francis, Inc., http://www.routledge-ny.com)

In any numerical solution, it is desirable for the numerical method to be unconditionally stable. Unconditional stability is especially important at the microscale where the mesh size needs to be extremely small for accuracy. This may cause problems in solution if the method is conditionally stable. Dai and Nassar (1999c) developed a two-level finite difference scheme of the Crank-Nicholson type by introducing an intermediate function for solving Eq. (6.3.7) in one dimension. It was shown by the discrete energy method that the scheme is unconditionally stable. Furthermore, the truncation error of the scheme is $0(\Delta t^2 + \Delta x^2)$. The method is described as follows.

Let $\theta(x,t) = \frac{T(x,t)-T_0}{T_1-T_0}$, $\beta = \frac{t}{2\tau_q}$, and $\delta = \frac{x}{\sqrt{\alpha\tau_q}}$. Ignoring the heat source, Eq. (6.3.7), can be expressed in the general form as

$$\frac{\partial^2 \theta}{\partial \delta^2} + B \frac{\partial^3 \theta}{\partial \delta^2 \partial \beta} = A \frac{\partial \theta}{\partial \beta} + D \frac{\partial^2 \theta}{\partial \beta^2}, \tag{6.4.1}$$

where $B = \frac{\tau_T}{2\tau_q}$, and A and D are positive constants. For developing the numerical scheme, Eq. (6.4.1) is written as

$$\frac{\partial^2}{\partial \delta^2} \left(\theta + B \frac{\partial \theta}{\partial \beta} \right) = \frac{\partial}{\partial \beta} \left(A\theta + D \frac{\partial \theta}{\partial \beta} \right). \tag{6.4.2}$$

Define an intermediate function u as

$$u = A\theta + D \frac{\partial \theta}{\partial \beta}. \tag{6.4.3}$$

Substituting $\frac{\partial \theta}{\partial \beta}$ from Eq. (6.4.3) into Eq. (6.4.2), one obtains, for the case $D - BA \geq 0$,

$$\frac{\partial^2}{\partial \delta^2} ((D - BA)\theta + Bu) = D \frac{\partial u}{\partial \beta}. \tag{6.4.4}$$

For the case $D - BA < 0$, one may substitute the value for θ from Eq. (6.4.3) into Eq. (6.4.2) to obtain

$$\frac{\partial^2}{\partial \delta^2} \left(u + (BA - D) \frac{\partial \theta}{\partial \beta} \right) = A \frac{\partial u}{\partial \beta}. \tag{6.4.5}$$

For a proof of stability, one needs to consider the sign of $D - BA$.

Equations (6.4.4) and (6.4.5) can now be discretized using a Crank-Nicholson type of finite difference to obtain a solution for given initial and boundary conditions. The solution is unconditionally stable and is second-order accurate.

A compact finite difference scheme which is unconditionally stable and with a truncation error of order $(\Delta t^2 + \Delta x^4)$ was developed by Dai and Nassar (2001a) for solving Eq. (6.3.7). If we let

$$\theta = T + \tau_q \frac{\partial T}{\partial t}, \tag{6.4.6}$$

and

$$f = \frac{\partial^2 T}{\partial x^2}, \tag{6.4.7}$$

then Eq. (6.3.7) can be expressed as

$$\frac{1}{\alpha} \frac{\partial \theta}{\partial t} = f + \tau_T \frac{\partial f}{\partial t} + g, \tag{6.4.8}$$

where $g = \frac{1}{k} \left(Q + \tau_q \frac{\partial Q}{\partial t} \right)$. Equations (6.4.6) and (6.4.8) are discretized using the trapezoidal rule, and Eq. (6.4.7) is discretized using a fourth-order compact finite difference. Higher order compact finite differences can be found in

Lele (1992). As such, the compact finite difference scheme for Eqs. (6.4.6)-(6.4.8) is given by

$$\frac{1}{\alpha \Delta t}(\theta_j^{n+1} - \theta_j^n) = \frac{1}{2}(f_j^{n+1} + f_j^n) + \frac{\tau_T}{\Delta t}(f_j^{n+1} - f_j^n) + g_j^{n+\frac{1}{2}}, \qquad (6.4.9)$$

$$\frac{1}{2}(\theta_j^{n+1} + \theta_j^n) = \frac{1}{2}(T_j^{n+1} + T_j^n) + \frac{\tau_q}{\Delta t}(T_j^{n+1} - T_j^n), \qquad (6.4.10)$$

and

$$\frac{1}{10}f_{j-1}^n + f_j^n + \frac{1}{10}f_{j+1}^n = \frac{6}{5\Delta x^2}(T_{j-1}^n - 2T_j^n + T_{j+1}^n), \qquad (6.4.11)$$

where $1 \leq j \leq N - 1$. Here, θ_j^n denotes $\theta(j\Delta x, n\Delta t)$, where Δx and Δt are the spatial and temporal mesh sizes, $N\Delta x = L$ (film thickness), and $n\Delta t = t$. Equations (6.4.9)-(6.4.11) can be solved for the temperature $T(x,t)$. It was shown using a simple numerical example that the error as a function of time was smaller for the compact finite-difference scheme than for the finite difference scheme in Dai and Nassar (1999c). Also, the error decreased with a decrease in the grid size Δx.

Derivation of the dual-phase-lag Eq. (6.3.7) was based on the Taylor series first-order approximation to Eq. (6.3.1). Therefore, it would be desirable to study thermal behavior at the microscale by solving the original coupled Eqs. (6.3.1) and (6.3.4). An approximate analytic method for solving these coupled equations in one dimension has been developed by Dai and Nassar (2002a). To explain the method, we consider first the case without a heat source. Eliminating q from Eqs. (6.3.1) and (6.3.4), one has a one-dimensional dual-phase-lag equation given by

$$\frac{\partial T(x, t + \tau_q)}{\partial t} = a\frac{\partial^2 T(x, t + \tau_T)}{\partial x^2}, \qquad (6.4.12)$$

where, $a = \frac{k}{C_p}$. The initial and boundary conditions are assumed to be

$$T(x, 0) = \varphi(x) \text{ and } T(0, t) = T(L, t) = 0. \qquad (6.4.13)$$

Employing the method of separation of variables, we let $T(x, t) = F(t)X(x)$ and substitute it into Eq. (6.4.12) to obtain

$$F'(t + \tau_q)X(x) = aF(t + \tau_T)X''(x). \qquad (6.4.14)$$

Separating the x and t variables in Eq. (6.4.14), one obtains

$$\frac{F'(t + \tau_q)}{aF(t + \tau_T)} = \frac{X''(x)}{X(x)} = -\lambda, \qquad (6.4.15)$$

where λ is a constant. As such, Eqs. (6.4.12) and (6.4.13) can be separated into two ordinary differential equations

$$X''(x) + \lambda X(x) = 0, \ X(0) = X(L) = 0, \tag{6.4.16}$$

and

$$F'(t + \tau_q) = -a\lambda F(t + \tau_T). \tag{6.4.17}$$

A solution to Eq. (6.4.12) with initial and boundary conditions in Eq. (6.4.13) can be obtained by solving Eqs. (6.4.16) and (6.4.17). It is seen that a solution to Eq. (6.4.16) is

$$X_n(x) = \sin\frac{n\pi x}{L}, \ \lambda_n = (\frac{n\pi}{L})^2, \ n = 1, 2, 3, ... \tag{6.4.18}$$

With λ_n known, Eq. (6.4.17) becomes

$$F'_n(t + \tau_q) = -a\lambda_n F_n(t + \tau_T), \tag{6.4.19}$$

with

$$F_n(0) = \frac{2}{L}\int_0^L \varphi(x)\sin\frac{n\pi x}{L}dx, \ n = 1, 2, 3, ... \tag{6.4.20}$$

The analytic solution to Eqs. (6.4.12) and (6.4.13) can be now written as

$$T(x, t) = \sum_{n=1}^{\infty} F_n(t)\sin\frac{n\pi x}{L}, \tag{6.4.21}$$

where $F_n(t)$ is the solution to Eqs. (6.4.19) and (6.4.20). This solution was obtained by approximating $F_n(t)$ by a polynomial

$$F_n(t) = \sum_{i=0}^{M} C_i^n t^i, \tag{6.4.22}$$

where M is an arbitrary large integer. Solutions to C_i^n when $\tau_T \geq \tau_q$ and when $\tau_T < \tau_q$ are given in Dai and Nassar (2002a).

Dai and Nassar (2000a, 2001b) extended the finite difference numerical method and the compact finite difference scheme for solving the dual-phase-lag model in Eq. (6.3.7) to the 3-D thin film, where the thickness is at microscale. For this case, one may consider a film in which lags occur only in the thickness direction. Hence, the heat flux components in the x and y directions satisfy the traditional Fourier law, while the component in the thickness or z direction satisfy the dual-phase-lag Eq. (6.3.6). As such, one has

$$q_1 = -k\frac{\partial T}{\partial x}, \tag{6.4.23}$$

$$q_2 = -k\frac{\partial T}{\partial y}, \tag{6.4.24}$$

and

$$q_3 + \tau_q \frac{\partial q_3}{\partial t} = -k \left[\frac{\partial T}{\partial z} + \tau_T \frac{\partial}{\partial t} \left(\frac{\partial T}{\partial z} \right) \right]. \tag{6.4.25}$$

Substituting Eqs. (6.4.23) and (6.4.24) into Eq. (6.3.4), one has

$$-\frac{\partial q_3}{\partial z} = C_p \frac{\partial T}{\partial t} - Q - k \frac{\partial^2 T}{\partial x2} - k \frac{\partial^2 T}{\partial y^2}. \tag{6.4.26}$$

Differentiating Eq. (6.4.25) with respect to z and combining it with Eq. (6.4.26), one obtains the 3-D dual-phase-lag expression

$$\frac{1}{\alpha} \frac{\partial T}{\partial t} = \nabla^2 T + \frac{\partial}{\partial t} \left(\tau_q \frac{\partial^2 T}{\partial x^2} + \tau_q \frac{\partial^2 T}{\partial y^2} + \tau_T \frac{\partial^2 T}{\partial z^2} \right) + \frac{1}{k} \left(Q + \tau_q \frac{\partial Q}{\partial t} \right) - \frac{\tau_q}{\alpha} \frac{\partial^2 T}{\partial t^2}. \tag{6.4.27}$$

Equation (6.4.27) can be expressed as

$$\frac{1}{\alpha} \frac{\partial \theta}{\partial t} = f_x + f_y + f_z + \tau_q \frac{\partial}{\partial t} (f_x + f_y) + \tau_T \frac{\partial f_z}{\partial t} + \frac{1}{k} \left(Q + \tau_q \frac{\partial Q}{\partial t} \right), \tag{6.4.28}$$

where $\theta = T + \tau_q \frac{\partial T}{\partial t}$, $f_x = \frac{\partial^2 T}{\partial x^2}$, $f_y = \frac{\partial^2 T}{\partial y^2}$, and $f_z = \frac{\partial^2 T}{\partial z^2}$. To obtain a solution, Eq. (6.4.28) and $\theta = T + \tau_q \frac{\partial T}{\partial t}$ were discretized using the trapezoidal method, and f_x, f_y, f_z were discretized using the fourth-order compact finite difference scheme. The solution using this approach was shown to be unconditionally stable and with truncation error $(\Delta t^2 + \Delta x^4 + \Delta y^4 + \Delta z^4)$ (Dai and Nassar 2000a). Equation (6.4.28) was solved also by introducing an intermediate function as discussed in Dai and Nassar (1999c) and employing the preconditioned Richardson iteration. It was shown by the discrete energy method that the solution was unconditionally stable. However, the solution is second-order accurate which is not as accurate as the compact finite difference solution (Dai and Nassar 2000a). The same finite difference scheme with an intermediate function was used by Dai and Nassar (2002b) to solve the three-dimensional dual-phase-lag equation where lags were considered in the x, y, and z directions. The 3-D heat transport equation was obtained by differentiating Eq. (6.3.6) with regard to x, y, and z and then substituting the results into Eq. (6.3.4). As such, one obtains

$$\frac{\partial T}{\partial t} + A \frac{\partial^2 T}{\partial t^2} = B \nabla^2 T + C \frac{\partial}{\partial t} \nabla^2 T + G, \tag{6.4.29}$$

where $A = \tau_q$, $B = \frac{k}{C_p}$, $C = \frac{k \tau_T}{C_p}$, and $G = \frac{1}{C_p} \left(Q + \tau_q \frac{\partial Q}{\partial t} \right)$. Here, the intermediate function is

$$u = T + A \frac{\partial T}{\partial t}. \tag{6.4.30}$$

We consider two cases, (1) $AB - C \geq 0$ ($\tau_q \geq \tau_T$) and (2) $AB - C < 0$ ($\tau_q < \tau_T$). For case 1, one substitutes $\frac{\partial T}{\partial t} = \frac{1}{A}(u - T)$ from Eq. (6.4.30) into Eq. (6.4.29) to obtain

$$\frac{\partial u}{\partial t} = \left(\frac{\partial^2}{\partial x^2} + \frac{\partial^2}{\partial y^2} + \frac{\partial^2}{\partial z^2}\right)\left(\left(B - \frac{C}{A}\right)T + \frac{C}{A}u\right) + G. \qquad (6.4.31)$$

For case 2, substituting $T = u - A\frac{\partial T}{\partial t}$ from Eq. (6.4.30) into Eq. (6.4.29), one has

$$\frac{\partial u}{\partial t} = \left(\frac{\partial^2}{\partial x^2} + \frac{\partial^2}{\partial y^2} + \frac{\partial^2}{\partial z^2}\right)\left(Bu + (C - AB)\frac{\partial T}{\partial t}\right) + G. \qquad (6.4.32)$$

Equations (6.4.31) and (6.4.32) were discretized using a Crank-Nicholson type finite difference and the preconditioned Richardson iteration employed to obtain a solution. The solution is second-order accurate and is unconditionally stable for the two cases above.

In further studies Dai and Nassar (2000b,c; 2001c) and Dai et al. (2001) developed numerical methods for solving the three-dimensional heat transport Eqs. (6.3.6) and (6.3.4) for two layers (gold and chromium) in the case of rectangular or arbitrary domains with dual lag in the z-direction. For a rectangular domain, a Crank-Nicholson finite difference scheme was used. The three-dimensional implicit scheme was solved using a preconditioned Richardson iteration. Applying a parallel Gaussian elimination procedure, a domain decomposition algorithm was developed to solve the two tridiagonal linear systems (one for each layer) with unknowns at the interface (Dai and Nassar 2000b). A similar procedure, applying the parallel Gaussian elimination coupled with Newton's iteration, was used to solve the same heat equations with nonlinear interfacial conditions (Dai and Nassar 2001c). A hybrid finite element-finite difference scheme was used to solve the same problem where the shape of the film can be cylindrical or arbitrary in the xy-directions and where the interfacial conditions are linear or nonlinear (Dai and Nassar 2000c; Dai et al. 2001).

6.5 Short-Pulse Laser Heating of Thin Films

Short-pulse laser heating of multi-layer metal films of gold and chromium using the electron-phonon two-step model in the case of temperature-dependent thermal conductivity and heat capacity was considered by Qiu and Tien (1994). The electron heat capacity in Eq. (6.2.2a) was assumed proportional to the electron temperature, and the thermal conductivity was proportional to the ratio of the electron temperature and the lattice temperature. As such, Eqs. (6.2.2a,b) become

$$C_e(T_e)\frac{\partial T_e}{\partial t} = \frac{\partial}{\partial x}\left(\frac{T_e}{T_l}\kappa\frac{\partial T_e}{\partial x}\right) - G(T_e - T_l) + Q \qquad (6.5.1a)$$

$$C_l\frac{\partial T_l}{\partial t} = G(T_e - T_l). \qquad (6.5.1b)$$

This model is termed the parabolic two-step model by Qiu and Tien (1994). The laser source Q, assuming a Gaussian beam, is given by .

$$Q = 0.94 \frac{1-R}{t_p \delta} J \exp\left[-\frac{x}{\delta} - 2.77 \left(\frac{t}{t_p} \right)^2 \right]. \qquad (6.5.2)$$

where δ is the penetration depth of laser radiation, R is reflectivity, J is laser fluence, and t_p is the laser pulse duration, full-width-at-half-maximum (FWHM).

The initial conditions are given by

$$T_e(x, -2t_p) = T_l(x, -2t_p) = T_0 = 300K, \qquad (6.5.3)$$

and the boundary conditions, neglecting heat losses from the front and back surfaces, by

$$\frac{\partial T_e}{\partial x}\Big|_{x=0} = \frac{\partial T_e}{\partial x}\Big|_{x=L} = \frac{\partial T_l}{\partial x}\Big|_{x=0} = \frac{\partial T_l}{\partial x}\Big|_{x=L} = 0. \qquad (6.5.4)$$

The boundary condition at the interface between gold (Au) and chromium (Cr), assuming continuity of flux and temperature, is expressed as

$$\frac{T_e}{T_l} \kappa \frac{\partial T_e}{\partial x}\Big|_{Au} = \frac{T_e}{T_l} \kappa \frac{\partial T_e}{\partial x}\Big|_{Cr}; \quad T_e\Big|_{Au} = T_e\Big|_{Cr}. \qquad (6.5.5)$$

A numerical solution was obtained by employing the semi-implicit Crank-Nicholson scheme. A uniform grid spacing with 400 grid points was used. The laser pulse duration was 0.1 ps, $R = 0.93$, $\delta = 15.3nm$, and the laser intensity was 500 Jm^{-2}. The gold and chromium thermal conductivities, lattice and electron heat capacities, and the electron-phonon coupling factor used in the numerical simulation were $\kappa_{Au} = 315Wm^{-1}K^{-1}$, $\kappa_{Cr} = 94Wm^{-1}K^{-1}$, $C_l = 2.5 \times 10^6 Jm^{-3}K^{-1}$ for gold, $C_l = 3.3 \times 10^6 Jm^{-3}K^{-1}$ for chromium, $C_e = 2.1 \times 10^4 Jm^{-3}K^{-1}$ for gold, $C_e = 5.8 \times 10^4 Jm^{-3}K^{-1}$ for chromium, $G_{Au} = 2.6 \times 10^{16}Wm^{-3}K^{-1}$, and $G_{Cr} = 42 \times 10^{16}Wm^{-3}K^{-1}$. Figure 6.5.1 shows the temperature profiles in a 1000 $\overset{o}{A}$ thick gold film. By comparison, Fig. (6.5.2) shows temperature profiles for a 500 $\overset{o}{A}$ thick gold film on top of a 500 $\overset{o}{A}$ thick chromium layer. It is seen that the electron temperature profile in the gold film at times 0.5 , 2, and 6 ps is considerably lower for the two-layer film, Fig. (6.5.2) than for the one-layer film in Fig. (6.5.1). Also, the temperature profile in the lattice of the gold film is higher for the single than for the two-layer film. It is seen that, for the two-layer film, the bulk of the temperature is in the chromium layer. These results are explained by the fact that chromium has a higher electron-lattice coupling factor (G) than gold. As a result, energy absorbed in the gold layer is transported by electrons to the chromium layer where it is converted into chromium lattice energy. As such, chromium acts as a heat sink which significantly reduces the temperature rise in the top gold layer. This suggests a new way to increase the resistance of gold-coated metals, such as mirrors used in high-power laser systems, to thermal damage. Further calculations of temperature profiles in a gold-chromium-gold triple layer film, showed that the sandwiched chromium layer, owing to its low thermal diffusivity, blocks the heat energy from being transported farther to the bottom gold layer. Thus, most of the heat is absorbed in the chromium layer leaving both gold layers unaffected by any high temperature rise.

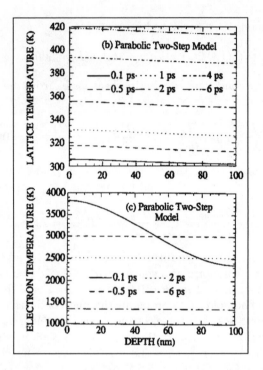

Fig. 6.5.1. Lattice and electron temperatures as functions of depth in a 1000 $\overset{O}{A}$ thick gold film (reprinted from Qiu and Tien 1994 with permission from Elsevier Science)

Figure (6.5.3) shows good agreement between predicted transient electron temperature profiles from the electron-phonon two-step model and measured results. It is seen that there is a faster surface temperature drop in the two-and three-layer films than in the single-layer gold film. On the other hand, there was no substantial difference between the two-and three-layer films (Qiu et al. 1994). Furthermore, comparisons of experimental results with predicted transient electron temperature changes over time, $\frac{\Delta T_e}{(\Delta T_e)_{\max}}$, for 200 $\overset{o}{A}$ and 1000 $\overset{o}{A}$ single-layer gold thicknesses, 13.4 $\frac{I}{m^2}$ and $t_p = 0.1ps$, revealed that the surface temperature drop over time was more rapid for the thick than for the thin film. This is explained by the fact that heat transport by electrons is limited in the thin film. On the other hand, the larger region in the thick film allows for energy transport by electrons away from the surface, resulting in the temperature drop.

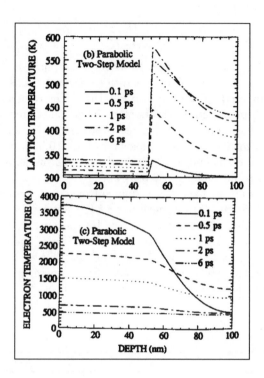

Fig. 6.5.2. Lattice and electron temperatures as functions of depth in a 500 Å gold/ 500 Å chromium two-layer film (reprinted from Qiu and Tien 1994 with permission from Elsevier Science)

As evidence for the lagging behavior in flux and in temperature gradient, Tzou (1995c) applied the lag model in Eq. (6.3.7) to the experimental data by Qiu et al. (1994) and Brorson et al. (1987) concerning surface temperature change over time in a gold thin film. Parameters used in the analysis were the same as those in Qiu et al. (1994) except for the heat source. The heat source Q in Eq. (6.5.2) was modified to give a better fit to the measured normalized autocorrelation function of the power intensity of the laser beam (Qiu et al. 1994). As such, Q is given by

$$Q = 0.94 \frac{1 - R}{t_p \delta} J \exp\left[-\frac{x}{\delta} - a\frac{|t - 2t_p|}{t_p}\right], \qquad (6.5.6)$$

where $a = 1.88$, and $t_p = 100 \ fs$. The initial conditions at time $t = 0$ are given by

$$T(x,0) = T_0 = 300K \text{ and } \frac{\partial T}{\partial t}(x,0) = 0. \qquad (6.5.7)$$

The boundary conditions at the front $(x = 0)$ and back $(x = L)$ surfaces of the gold film are given by

$$\frac{\partial T}{\partial x}(0,t) = \frac{\partial T}{\partial x}(L,t) = 0. \qquad (6.5.8)$$

The solution to Eq. (6.3.7) (referred to as the dual-phase lag, DPL, model), with source, initial, and boundary conditions given by Eqs. (6.5.6)-(6.5.8), was determined by the Laplace transform method. The Riemann-sum approximation was used for obtaining the Laplace inversion. Comparisons between predicted normalized temperature profiles over time at the surface of the film $(x = 0)$ and experimental results are shown in Figs. (6.5.4) and (6.5.5) for 0.1 μm and 0.2 μm film thicknesses. It is seen that the model agrees well with the experimental results for $\tau_q = 8.5$ ps and $\tau_T = 90$ ps. Thickness beyond 0.1 μm does not seem to have any significant effect on the lags because the same lags seem to fit the data in both figures (although to a somewhat lesser extent in Fig. (6.5.5)). As was pointed out by the author, the lags may be substantially affected by the thin structure of the film if one were to consider thicknesses significantly less than 0.1 μm. The fact that $\tau_T > \tau_q$ in the above analysis indicates that the heat flux precedes the temperature gradient or that flux is the cause and temperature gradient is the effect.

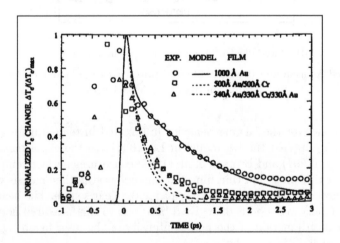

Fig. 6.5.3. Comparisons of predicted electron temperature changes over time from the electron-phonon model with experimental results (reprinted from Qiu et al. 1994 with permission from Elsevier Science)

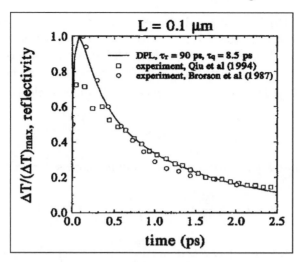

Fig. 6.5.4. Comparisons of predicted normalized electron temperature changes over time at the front surface of a 0.1 μm thick gold film from the dual-phase lag (DPL) model with experimental results (reprinted from Tzou 1995c with permission from the American Institute of Aeronautics and Astronautics, Inc., the American Physical Society, and Elsevier Science)

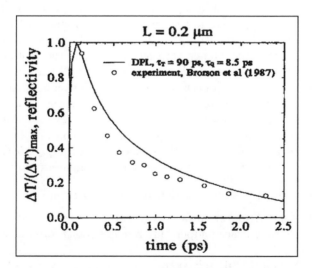

Fig. 6.5.5. Comparisons of predicted normalized electron temperature changes over time at the front surface of a 0.2 μm thick gold film from the dual-phase lag (DPL) model with experimental results (reprinted from Tzou 1995c with permission from the American Institute of Aeronautics and Astronautics, Inc. and the American Physical Society)

Lin et al. (1997) investigated the effects of phase lags and time on the temperature profiles in the thickness direction of a thin metal film using the one-dimensional dual-phase-lag equation. The following dimensionless variables were introduced: $\theta = \frac{T-T_0}{T_w-T_0}$, $\delta = \frac{x}{L}$, $\beta = \frac{t}{(\frac{L^2}{\alpha})}$, $z_T = \frac{\alpha\tau_T}{L^2}$, and $z_q = \frac{\alpha\tau_q}{L^2}$. As such, the phase lag Eq. (6.3.7) , with $Q = 0$, in dimensionless form becomes

$$\frac{\partial^2\theta}{\partial\delta^2} + z_T\frac{\partial^3\theta}{\partial\delta^2\partial\beta} = \frac{\partial\theta}{\partial\beta} + z_q\frac{\partial^2\theta}{\partial\beta^2}. \tag{6.5.9}$$

The initial and boundary conditions in dimensionless form are given by

$$\theta = 0 \text{ and } \frac{\partial\theta}{\partial\beta} = \dot{T_0}\frac{L^2}{\alpha(T_w - T_0)} \text{ at } \beta = 0, \tag{6.5.10}$$

$$\theta = 1 \text{ at } \delta = 0 \text{ and } \frac{\partial\theta}{\partial\delta} = 0 \text{ at } \delta = 1. \tag{6.5.11}$$

Here, T_0 is the initial temperature in the solid, T_w is a step temperature input (heat source) at the top film surface ($x = 0$), and $\dot{T_0}$ is an initial rate of change of temperature with time. An exact solution of Eq. (6.5.9) was obtained using the separation of variables technique. Figure (6.5.6) shows temperature distributions as a function of δ for different lags (τ_T), for $\beta = 0.05$, and for $z_q = 0.05$. It is seen that the temperature in the top region of the film ($\delta \leq 0.2$) decreases with an increase in the lag z_T. Also, when $z_T = 0$, the dual-phase-lag equation reduces, as expected, to the wave equation. Figure (6.5.7) shows the effect of time on the temperature distributions in the thickness direction of the film. It is seen that the temperature, as a function of the depth δ, increases with an increase in the dimensionless time β. As is well known, the limiting temperature distribution approaches that of the diffusion heat equation.

A similar analysis by Antaki (1998) using the Fourier and Laplace double-transforms solution gave results of temperature profiles similar to those in Fig. (6.5.6). For this analysis, it was assumed that a constant heat flux q_0 is applied suddenly and uniformly on the surface ($x = 0$). Fig. (6.5.8) shows the effect of changing $\frac{\tau_T}{\tau_q}$ ($B = \frac{1}{2}\frac{\tau_T}{\tau_q}$) on the surface temperature of the film as a function of time. It is seen that the surface temperature $\left[\varphi = \frac{k(T-T_0)}{(q_0(\alpha\tau_q)^{\frac{1}{2}})}\right]$ increases with time. Also, the surface temperature decreased with an increase in B. As a result, the surface temperature remains cooler for films with a large τ_T relative to τ_q.

Chen and Beraun (2001) solved the two-step model in Eqs. (6.2.2a,b) by modifying the heat flux vector in Eq. (6.2.2a) and by introducing heat conduction into Eq. (6.2.2b) and compared solutions to determine the effect of heat conduction in the metal lattice. Equations (6.2.2a,b) are expressed now as

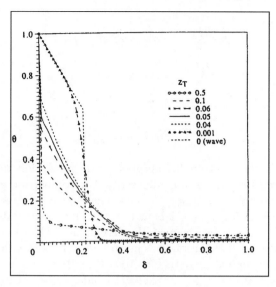

Fig. 6.5.6. Temperature distributions as a function of δ (film thickness) for different lags (τ_T), for β =0.05, and for z_q=0.05 (reprinted from Lin et al. 1997 with permission from Elsevier Science)

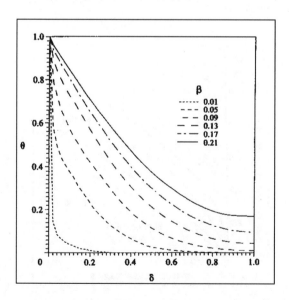

Fig. 6.5.7. Temperature distributions in the thickness direction of the film for different times (β) and for $z_q = z_T = 0.05$ (reprinted from Lin et al. 1997 with permission from Elsevier Science)

$$C_e(T_e)\frac{\partial T_e}{\partial t} = -\nabla q_e - G(T_e - T_l) + Q, \qquad (6.5.12)$$

$$C_l\frac{\partial T_l}{\partial t} = -\nabla q_l + G(T_e - T_l), \qquad (6.5.13)$$

where the flux vector $q_{e(l)}$ is given by

$$\tau_{e(l)}\frac{\partial q_{e(l)}}{\partial t} + q_{e(l)} = -\kappa_{e(l)}\nabla T_{e(l)}. \qquad (6.5.14)$$

Here, $\tau_{e(l)}$ is the electron (e) or phonon (l) relaxation time, and $\kappa_{e(l)}$ is the electron (lattice) thermal conductivity. For comparison purposes, Eqs. (6.2.2a,b) were referred to as the PA model. The same equations with the heat flux given by Eq. (6.5.14) were labelled as the HQ model. Also, the model in Eqs. (6.5.12)-(6.5.14) was named the HH model. The laser heat source, assuming a Gaussian distribution, was given by Eq. (6.5.2) with t replaced by $t - 2t_p$. The initial conditions were $T_e(x,0) = T_l(x,0) = T_0$, and the boundary conditions were $q_e(0,t) = q_e(L,t) = q_l(0,t) = q_l(L,t) = 0$. The temperature-dependent thermal conductivity and heat capacity of electrons are given by $\kappa_e = \kappa_0\frac{T_e}{T_l}$ and $C_e = A_eT_e$.

Here, the material constants for gold are given by $\kappa_0 = 315Wm^{-1}K^{-1}$, $A_e = 70\ Jm^{-3}K^{-2}$, $C_l = 2.5 \times 10^6\ Jm^{-3}K^{-1}$, $\kappa_l = 315Wm^{-1}K^{-1}$, $G = 2.6 \times 10^{16}\ Wm^{-3}K^{-1}$, $\tau_e = 0.04ps$, and $\tau_l = 0.8ps$. The equations for the PA, HQ, and HH models were solved numerically using the Corrective Smooth Particle Method (CSPM).

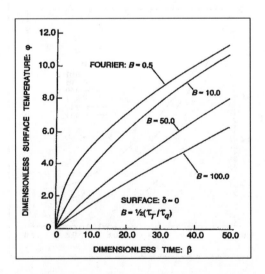

Fig. 6.5.8. Surface temperature distributions over time in a semi-infinite slab for different values of B $=\frac{1}{2}\frac{\tau_T}{\tau_q}$ (reprinted from Antaki 1998 with permission from Elsevier Science)

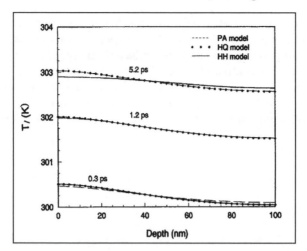

Fig. 6.5.9. Lattice temperature as a function of depth calculated from the PA, HQ, and HH models in the text for a 100 nm gold film irradiated with a 0.1 ps laser pulse with a fluence of $\frac{10\mathrm{J}}{\mathrm{m}^2}$ (reprinted from Chen and Beraun 2001 with permission from Taylor & Francis, Inc., http://www.routledge-ny.com)

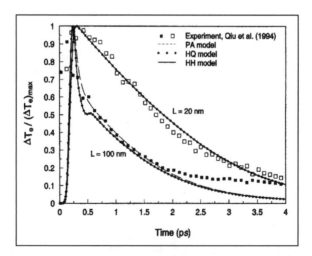

Fig. 6.5.10. Comparisons between calculated and experimental observations of the normalized electron temperatures at the front surface of 20nm and 100nm gold films irradiated with a 0.1ps laser pulse at $\frac{13.4\mathrm{J}}{\mathrm{m}^2}$ (reprinted from Chen and Beraun 2001 with permission from Taylor & Francis,Inc., http://www.routledge-ny.com)

Details of this method are in Chen et al. (1999). Differences in results among the three models, concerning the lattice temperature as a function of depth, were not substantial as seen in Fig. (6.5.9). Figure (6.5.10) compares the three models with regard to their fit to experimental data on the normalized electron temperatures at the front surface of the gold film. It is seen again that all three models compared favorably. Also, the analysis revealed that the first term in Eq. (6.5.14) concerning the metal lattice did not have any significant effects on results. One would expect that the effect of conductivity in the lattice would gain in importance as the film increases in thickness.

References

1. Abakians H, Modest MF (1988) Evaporative cutting of a semitransparent body with a moving CW laser. J Heat Transfer 110: 924-930
2. Achilias DS, Kiarissides C (1992) Development of a general mathematical framework for modeling diffusion-controllde free-radical polmerization reactions. Macromolecules 25: 3739-3750
3. Afanasiev YV, Isakov VA, Zavestovskaya LN, Chichkov BN, von Alvensleben F, Welling H (1997) Applied Physics A 64: 561-572
4. Al-Nimr MA, Arpaci VS (2001) The thermal behavior of thin films in the hyperbolic two-step model. Int J Heat and Mass Transfer 43: 2021-2028
5. Ameel TA, Warrington RO, Yu D, Dahlbacka G. (1994) Thermal analysis of x-ray irradiated thick resists to determine induced structurer deformations. Tenth International Heat Transfer Conference, Brighton, England, Heat Transfer 5: 313-319, Ed. GF Hewitt
6. Anisimov SI, Kapeliovich BL, Perelman TL (1974) Electron emission from metal surfaces exposed to ultra-short laser pulses. Soviet Physics JETP 39: 375-377
7. Anseth KS, Bowman CN (1993) Reaction diffusion enhanced termination in polymerizations of multifunctional monomers. Polymer Reaction Engineering 1: 499-520
8. Anseth KS, Wang CM, Bowman CN (1994) Reaction behavior and kinetic constants for photopolymerizations of multi(meth)acrylate monomers. Polymer 35: 3243-3250
9. Antaki PJ (1998) Solution for non-Fourier dual phase lag heat conduction in a semi-infinite slab with surface heat flux. Int J Heat Mass Transfer 41: 2253-2258
10. Arnold N, Baeuerle D (1993b) Simulation of growth in pyrolytic laser-CVD of microstructures II. two-dimensional approach. Microelectronic Engineering 20: 43-54
11. Arnold N, Bityurin N (1999) Model for laser-induced thermal degradation and ablation of polymers. Appl Phys A 68: 615-625
12. Arnold N, Kullmer R, Baeuerle D (1993a) Simulation of growth in pyrolytic laser-CVD of microstructures I. one-dimensional approach. Microelctronic Engineering 20: 31-41
13. Arnold N, Thor E, Kirichenko N, Baeuerle D (1996) Pyrolytic LCVD of fibers: a theoretical description. Appl Phys A 62: 503-508

14. Baeuerle D (1986) Chemical processing with lasers, vol 1. Springer Series in Materials Science, Springer, Berlin
15. Baeuerle D (2000) Laser processing and chemistry, 3d edn. Springer Verlag, New York
16. Baeuerle D, Luk'yanchuk B, Piglmayer K (1990) On the reaction kinetics in laser-induced pyrolytic chemical processing. Appl Phys A 50: 385-396
17. Ballantyne A, Hyman H, Dym CL, Southworth R (1985) Response of lithography mask structures of intense repetitively pulsed x-rays: thermal stress analysis. J Appl Phys 58: 4717-4725
18. Banerjee I, Livengood RH (1993) Applications of ficused ion beam. J Electrochem Soc 140: 183-188
19. Bang SY, Modest MF (1991) Multiple reflection effects on evaporative cutting with a moving CW laser. J Heat Transfer 113: 663-669
20. Bang SY, Roy S, Modest MF (1993) CW laser machining of hard ceramics II. Effects of multiple reflections. Int J Heat Mass Transfer 36: 3529-3540
21. Barber DJ, Frank FC, Moss M, Steeds JW, Tong IST (1973) Prediction of ion-bombarded surface topographies using Frank's kinematic theory of crystal dissolution. J Mater Sci 8: 1030-1040
22. Barron RF (1985) Cryogenic Systems, 2nd edn. Oxford University Press, New York
23. Baum TH, Comita PB (1992) Laser-induced chemical vapor deposition of metals for microelectronic technology. Thin Solid Films 218: 80-94
24. Bell L, Swanson L (1985) Mechanisms of liquid metal ion source operation. Nuclear Instruments and Methods in Physics Research B 10-11: 783-787
25. Bityurin N, Muraviov S, Alexandrov A, Bronnikova N, Malyshev A (1997) UV laser modifications and etching of thin polymer films in different environments. SPIE 3093: 108-116
26. Bityurin N, Malyshev A (1998) UV-laser ablation of absorbing dielectrics by ultra-short laser pulses. Applied Surface Science 127-129: 199-205
27. Bityurin N, Arnold N, Luk'yanchuk B, Baeuerle D (1998) Bulk model of laser abaltion of polymers. Applied Surface Science 127-129: 164-170
28. Bityurin N (1999) UV etching accompanied by modifications. Surface etching. Applied Surface Science138-139: 354-358
29. Blauner P, Mauer J (1993) X-ray mask repair. IBM J Res Dev 37: 4721-4725
30. Bowman CN, Peppas, NA (1991) Coupling of kinetics and volume relaxation during polymerizations of multiacrylates and multimethacrylates. Macromolecules 24: 1914-1920
31. Brandup J, Immergut EH (1989) Polymer handbook, 3rd edn. John Wiley, New York, pp 77-79
32. Brannon JH, Scholl D, Kay E (1991) Ultraviolet photoablation of a plasmasynthesized fluorocarbon polymer. Appl Phys A 52: 160-166
33. Brorson SD, Fujimoto JG, Ippen, EP (1987) Femtosecond electron heattransport dynamics in thin gold film. Physical Review Letters 59: 1962-1965
34. Bryzak J, Petersen K, McCulley W (1994) Micromachines on the march. IEEE Spectrum 31: 20-31
35. Burns FC, Cain SR (1996) The effect of pulse repetition rate on laser ablation of polyimide and polymethylmethacrylate-based polymers. J Phys D: Appl Phys 29: 1349-1355
36. Cain SR (1993) A photothermal model for polymer ablation: Chemical modification. J Phys Chem 97: 7572-7577

37. Cain SR, Burns FC, Otis CE (1992a) On single-photon ultraviolet ablation of polymeric materials. J Appl Phys 71: 4107-4117
38. Cain SR, Burns FC, Otis CE, Braren B (1992b) Photothermal description of polymer ablation: Absorption behavior and degradation time scales. J Appl Phys 72: 5172-5178
39. Canuto C, Hussaini MY, Quarteroni A, Zang TA (1988) Spectral Methods in Fluid Dynamics, Springer-Verlag, New York, NY
40. Carslaw HS, Jaeger JC (1959) Conduction of Heat in Solids, 2nd ed., Clarendon Press, Oxford
41. Carter G (2001) The physics and applications of ion beam erosion. J of Physics D: Appl. Physics 34: 1-22
42. Carter G, Colligon JS, Nobes MJ (1971) The equilibrium topography of sputtered amorphous solids II. J Mater Sci 6: 115-117
43. Carter G, Colligon JS, Nobes MJ (1973) The growth of topography during sputtering of amorphous solids. J Mater Sci 8: 1473-1481
44. Carter G, Nobes MJ (1984) Theory of development of surface topography under spatiotemporally heterogeneous sputtering conditions. Nuclear Instruments and Methods in Physics Research 82: 635-639
45. Catana C, Colligon JS, Carter G (1972) The equilibrium topography of sputtered amorphous solids III. Computer Simulation. J Mater Sci 7: 467-471
46. Cattaneo C (1958) A form of heat conduction equation which eliminates the paradox of instantaneous propagation. Compute Rendus 247: 431-433
47. Chaker M, La Fontaine B, Cote CY, Kieffer JC, Pepin H, Talon MH, Enright GD, Villeneuve DM (1992) Laser plasma sources for proximity printing x-ray lithography. J Vac Sci Technol, B10: 3239-3242
48. Chandrupatla TR, Belegundu AD (1991) Introduction to finite elements in engineering. Prentice Hall, Englewood Cliffs, NJ, pp 334-371
49. Chan CL, Mazumder J (1987) One-dimensional steady-state model for damage by vaporization and liquid expulsion due to laser-material interaction. J Appl Phys 62: 4579-4586
50. Chen JK, Beraun JE (2001) Numerical study of ultrashort laser pulse interactions with metal films. Numerical Heat Transfer, A 40: 1-20
51. Chen JK, Beraun JE, Carney TC (1999) A corrective smoothed particle method for boundary value problems in heat conduction. Int J Numer Method Eng 46: 231-252
52. Chester M (1963) Second sound in solids. Physical Review 131: 2013-2015
53. Chiba A, Okada K (1990) Dynamic thermal distortion in an x-ray mask membrane during pulsed x-ray exposure. Japanese Journal of Applied Physics 29: 2610-2615
54. Chiba A, Okada K (1991) Dynamic in-plane thermal distortion analysis of an x-ray mask membrane for synchrotron radiation lithography. J Vac Sci Technol B 9: 3275-3279
55. Conde O, Kar A, Mazumder J (1992) Laser chemical vapor deposition of TiN dots: A comparison of theoretical and experimental results. J Appl Phys 72: 754-761
56. Cook WD (1991) Kinetics and properties of a photopolymerized dimethacrylate oligomer. J of Applied Polymer Science 42: 2209-2222
57. Costela A, Figuera JM, Florido F, Garcia-Moreno J, Collar EP, Sastre R (1995) Ablation of poly(methyl methacrylate) and poly(2-hydroxyethylmethacrylate) by 308, 222, and 193 nm excimer-laser radiation. Appl Phys A 60: 261-270

58. Crell A, Friedrich S, Schreiber HU, Weick A (1997) Focused ion-beam implanted lateral field-effect transistors on bulk silicon. J Appl Phys 82: 4616-4620

59. Crow G, Puretz J, Orloff J, DeFreez RK, Elliott RA (1988) The use of vector scanning for producing arbitrary surface contours with focused ion beam. J Vac Sci Technol B 6: 1605-1607

60. Dai W, Nassar R (1997a) Three-dimensional numerical model for thermal analysis in x-ray irradiated photoresists. Numerical Heat Transfer, Part A 31: 585-595

61. Dai W, Nassar R (1997b) Three-dimensional numerical model for thermal analysis in x-ray irradiated photoresists with cylindrical domain. Numerical Heat Transfer, A 32: 517-530

62. Dai W, Nassar R, Warrington RO, Shen B (1997) Three-dimensional numerical model for thermal analysis in x-ray irradiated photoresists. Numerical Heat Transfer, A 31: 585-595

63. Dai W, Nassar R (1998a) Preconditioned Richardson numerical method for thermal analysis in x-ray lithography with cylindrical domain. Numerical Heat Transfer, A 34: 599-616

64. Dai W, Nassar R (1998b) A three-dimensional numerical method for thermal analysis in x-ray lithography. International Journal of Numerical Methods for Heat and Fluid Flow 8: 409-423

65. Dai W, Nassar R (1999a) A hybrid finite element finite difference method for thermal analysis in x-ray lithography. International Journal of Numerical Methods for Heat and Fluid Fliow 9: 660-676

66. Dai W, Nassar R, Zhang C, Shabanian S, Maxwell J (1999b):A numerical model for simulating axisymmetric rod growth in three-dimensional laser chemical vapor deposition. Numerical Heat Transfer, A 36: 251-262

67. Dai W, Nassar R (1999c) A finite difference scheme for solving the heat transport equation at the microscale. Numerical Methods of Partial Differential Equations 15: 697-708

68. Dai W, Nassar R (2000a) A compact finite difference schemes for solving a three-dimensional heat transport equation in a thin film. Numer Methods Partial Differential Equations 16: 441-458

69. Dai W, Nassar R (2000b) A domain decomposition method for solving three-dimensional heat transport equations in a double-layered thin film with microscale thickness. Numerical Heat Transfer, A 38: 243-255

70. Dai W, Nassar R (2000c) A hybrid finite element-finite difference method for solving three-dimensional heat transport equations in a double-layered thin film with microscale thickness. Numerical Heat Transfer, A 38: 573-588

71. Dai W, Nassar R (2001a) A compact finite-difference scheme for solving a one-dimensional heat transport equation at the microscale. J. of Computational and Applied Mathematics 132: 431-441

72. Dai W, Nassar R (2001b) A finite difference scheme for solving a three-dimensional heat transport equation in a thin film with microscale thickness. Int. J. Numer. Meth. Engng. 50: 1665-1680

73. Dai W, Nassar R (2001c) A finite difference method for solving 3-D heat transport equations in a double-layered thin film with microscale thickness and non-linear interfacial conditions. Numerical Heat Transfer, A 39: 21-33

74. Dai W, Nassar R, Mo L (2001d) A domain decomposition method for solving 3-D heat transport equations in a double-layered cylindrical thin film with submicroscale thickness and nonlinear interfacial conditions. Numerical Heat Transfer, A 40: 619-638

75. Dai W, Nassar R (2002a) An approximate analytic method for solving 1D dual-phase-lagging heat transport equations. Int J Heat and Mass Transfer 45: 1585-1593

76. Dai W, Nassar R (2002b) An unconditionally stable finite difference scheme for solving a 3D heat transport equation in a sub-microscale thin film. J Computational and Applied Mathematics 145: 247-260

77. D'Couto GC, Babu SV (1994) Heat transfer and material removal in pulsed excimer-laser-induced ablation: pulsewidth dependence. J Appl Phys 76: 3052-3058

78. Decker C, Elzaouk B, Decker D (1996) Kinetic study of ultrafast photopolymerization reactions. JMS-Pure Appl Chem A 33: 173-190

79. Decker C (1987) UV-Curing chemistry: Past, present, and future. Journal of Coating Technology 59: 97-106

80. Deutsch TF, Geis MW (1983) Self-developing UV photoresist using excimer laser exposure. J Appl Phys 54: 7201-7204

81. Deville MO, Mund, EH, Van Kemenada V (1994) Preconditioned Chebyshev collocation methods and triangular finite elements. Comput Methods Appl Mech Engrg, 116: 193-200

82. Douglas J (1962) Alternating direction methods for three space variables. Numer Math 4: 41-63

83. Ducommun JP, Cantagrel M, Marchal M (1974) Development of a general surface contour by ion erosion: Theory and computer simulation. J Mater Sci 9: 725-736

84. Elsayed-Ali HE (1991) Femtosecond thermoreflectivity and thermotransmissivity of polycrystalline and single-crystalline gold films. Physical Review B 43: 4488-449

85. Feiertag G, Ehrfeld W, Lehr H, Schmidt A, Schmidt M (1997) Calculation and experimental determination of the structure transfer accuracy in deep x-ray lithography. J. Micromech. Microeng. 7: 323-331.

86. Flach L, Chartoff RP (1995a) A process model for nonisothermal photopolymerization with a laser light source I. Basic model development. Polymer Engineering and Science 35: 483-492

87. Flach L, Chartoff RP (1995b) A process model for nonisothermal photopolymerization with a laser light source II. Behavior in the vicinity of a moving exposed region. Polymer Engineering and Science 35: 493-498

88. Flik MI, Choi BI, Goodson KE (1991) Heat transfer regimes in microstructures. ASME DSC-32: 31-47

89. Frank FC (1958) Growth and perfection of crystals. John Wiley and Sons, New York

90. Furzikov NP (1990) Approximate theory of highly absorbing polymer abaltion by nanosecond laser pulses. Appl Physics Lett 56: 1638-1640

91. Gamo K(1993) Focused ion beam technology. Semiconductor Science and Technology 8: 1118-1123

92. Garrett RW, Hill DJT, O'Donnell JH, Pomery PJ, Winzor CL (1989) Application of ESR spectroscopy to the kinetics of free radical polymerization of methyl methacrylate in bulk to high conversion. Polymer Bulletin 22: 611-616

93. Garrison BJ, Srinivasan R (1984) Microscopic model for the ablative photodecomposition of polymers by far-ultraviolet radiation. Appl Phys Lett 44: 849-851

94. Gerald CF (1978) Applied Numerical Analysis. Addison Wesley, New York

95. Giannuzzi LA, Stevie FA (1999) A review of focused ion beam milling techniques for TEM specimen preparation. Micron 30: 197-204

96. Griffiths SK, Hruby JM, Ting A (1999) The influence of feature sidewall tolerance on minimum absorber thickness for LIGA x-ray masks: J. Micromech. Microeng. 9: 353-361

97. Griffiths SK, Ting A (2001) The influence of x-ray fluorescence on LIGA sidewall tolerances. Fourth International Workshop on High-Aspect-Ratio-Structure Technology, Forschungszentrum Karlsruhe GmbH, Ed., Karlsruhe, Germany: 161-162

98. Griffiths SK, Ting A (2002) The influence of x-ray fluorescence on LIGA sidewall tolerances. Microsystem Technologies 8: 120-128

99. Gross M, Harriott L, Opila R (1990) Focused ion beam simulated deposition of aluminum from trialkylamine alanes. J Appl Phys 68: 4820-4824

100. Guyer RA, Krumhansl JA (1966) Solution of the linearized Boltzmann equation. Physical Review 148: 766-778

101. Hamaguchi S, Dalvie M, Farouki RT, Sethuraman S (1993) A shock-tracking algorithm for surface evolution under reactive-ion etching. J Appl Phys 74: 5172-5184

102. Han J, Jensen KF (1994) Combined experimental and modeling studies of laser-assisted chemical vapor deposition of copper from copper(I)-hexafluoroacetylacetonate-trimethylvinylsilane. J Appl Phys 75: 2240-2250

103. Harriott LR (1993) Digital scan model for focused ion beam induced gas etching. J Vac Sci Technol B 11: 2012-2015

104. Hegab H, Shen B, Dai W, Ameel T (1998) Transient three-dimensional numerical model for thermal analysis in x-ray lithography. Numerical Heat Transfer, A 34: 805-819

105. Heinrich K, Betz H, Heuberger A, Pongratz S (1981) Computer simulation of resist profiles in x-ray lithography. J. Vac. Sci. Technol. 19: 1254-1258.

106. Heinrich K, Betz H, Heuberger A (1983) Heating and temperature-induced distortions of silicon x-ray masks. J Vac Sci Technol B 1: 1352-1357

107. Himmelbauer M, Bityurin N, Luk'yanshuk B, Arnold N, Baeuerle D (1997) UV-laser-induced polymer ablation: The role of volatile species. SPIE 3093: 220-224

108. Hoyle CE, Sundell PE, Trapp M, Kang DM, Sheng D, Nagarajan R (1991) Polymerization kinetics of mono-and multifunctional monomers initiated by high intensity laser pulses: dependence of rate on peak pulse intensity and chemical structure. SPIE 1559: 202-213

109. Huang X (1996) Fabrication of three-dimensional microstructures by UV laser-induced polymerization. MS Thesis, Louisiana Tech University

110. Hutchison RA, Richards JR, Aronson MT (1994) Determination of propagation rate coefficients by pulsed laser plymerization for sytems with rapid chain growth: vinyl acetate. Macromolecules 27: 4530-4537

111. Ishitani T, Ohnishi T, Kawanami Y (1990) Micromachining and device transplantation using focused ion beams. Jpn J Appl Phys 29: 2283-2287

112. Jellinek HHG, Srinivasan R (1984) Theory of etching of polymers by far-ultraviolet, high-intensity pulsed laser and long-term irradiation. J Phys Chem 88: 3048-3051

113. Joseph DD, Preziosi L (1989) Heat Waves. Reviews of Modern Physics 61: 41-73

114. Joshi AA, Majumbar A (1993) Transient ballistic and diffusive phonon heat transport in thin films. J Appl Phys 74: 31-39

115. Kaganov MI, Lifshitz IM, Tanatarov MV (1957) Relaxation between electrons and crystalline lattices. Soviet Physics JETP 4: 173-178

116. Kalburge A, Konkar A, Ramachandran T, Chen P, Madhukar A (1977) Focused ion beam assisted chemically etched mesas on GaAs(001) and the nature of subsequent molecular beam epitaxial growth. J Appl Phys 82: 859-864

117. Kar A, Rockstroh T, Mazumder J (1992) Two-dimensional model for laser-induced materials damage: Effects of assist gas and multiple reflections inside the cavity.J Appl Phys 71: 2560-2569

118. Keyes T, Clarke RH, Isner JM (1985) Theory of photoablation and its implications for laser phototherapy. J Phys Chem 89: 4194-4196

119. Ki H, Mohanty PS, Mazumder J (2001) Modelling of high-density laser-material interaction using fast level set method. J Phys D: Appl Phys 34: 364-372

120. Ki H, Mohanty PS, Mazumder J (2002) Modeling of laser keyhole welding: Part I. Mathematical modeling, numerical methodology, role of recoil pressure, multiple reflections, and free surface evolution. Metallurgical and Materials Transaction 33A: 1817-1830

121. Ki H, Mohanty PS, Mazumder J (2002) Modeling of laser keyhole welding: Part II. Simulation of keyhole evolution, velocity, temperature profile, and experimental verification. Metallurgical and Materials Transaction 33A: 1831-1842

122. Kirichenko N, Baeuerle D (1992) The influence of heterogenous and homogeneous reactions in laser-chemical processing. Thin Solid Films 218:1-7

123. Kirichenko N, Piglmayer NK, Baeuerle D (1990) On the kinetics of non-equimolecular reactions in laser chemical processing. Applied Physics A 51: 498-507

124. Koenig H, Mais N, Hofling E, Reithmaier J, Forchel A, Mussig H, Brugger H (1998) Focused ion beam implantation for opto- and microelectronic devices. J Vac Sci Technol B 16: 2562-2566

125. Konstantinov L, Nowak R, Hess P (1990) Gas-phase transport and kinetics in pulsed-laser CVD. Applied Surface Science 46: 102-107

126. Krueger R (1999) Dual-column (FIB-SEM) wafer applications. Micron 30: 221-226

127. Kueper S, Stuke M (1989) Ablation of polytetrafluoroethylene (Teflon) with femtosecond UV excimer laser pulses. Appl Phys Lett 54: 4-6

128. Kueper S, Stuke M (1987) Femtosecond UV excimer laser ablation. Appl Phys B 44: 199-204

129. Kueper S, Brannon J, Brannon K (1993) Threshold behavior in polyimide photoablation: Single-shot rate measurements and surface-temperature modeling. Appl Phys A 56: 43-50

130. Kurdikar L, Peppas N (1994) A kinetic model for diffusion-controlled bulk cross-linking photopolymerization. Macrmolecules 27: 4084-4092

131. Lakhsasi A, Pepin H, Skorek A (1997) Transient thermal stability of the x-ray mask SiC-W under short pulse irradiation. Simulation Practice and Theory 5: 315-331

132. Lazare S, Granier V (1989) Ultraviolet laser photoablation of polymers: A review and recent results. Laser Chem 10: 25-40

133. Lazare S, Guan W, Drilhole D (1996) High sensitivity quadrupole mass spectrometry of neutrals sputtered by UV-laser ablation of polymers. Applied Surface Science 96-98: 605-610

134. Lecamp L, Youssef B, Bunel C, Lebaudy P (1999) Photoinitiation polymerization of a dimethacrylate oligomer: 2. kinetic studies, Polymer 40: 1403-1409

135. Lele SK (1992) Compact finite difference schemes with spectral-like resolution. J Comput Phys 103: 16-42

136. Li DC, Chen JT, Chyuan SW, Sun CY (1996) Computer simulations for mask structure heating in x-ray lithography. Computers and Structures 58: 825-834

137. Li D, Chyuan S, Chen J, Sun C (1995) Thermomechanical response analysis of lithographic mask structure using finite element method. JSME Int J series A 38: 563-571

138. Lin CK, Hwang CH, Chang YP (1997) The unsteady solutions of a unified heat conduction equation. Int J Heat Mass Transfer 40: 1716-1719

139. Linfoot EH (1955) Recent advances in optics. Oxford, Clarendon Press.

140. Luk'yanchuk B, Piglmayer K, Kirichenko N, Baeuerle D (1992) Inversion effects in the kinetics of laser-chemical processing. Physica A 180: 285-294

141. Luk'yanchuk B, Bityurin N, Anisimov S, Baeuerle D (1993a) The role of excited species in UV-laser materials ablation. Part I: Photophysical ablation of organic polymers. Appl Phys A 57: 367-374

142. Luk'yanchuk B, Bityurin N, Anisimov S, Baeuerle D (1993b) The role of excited species in UV-laser materials ablation. Part II: The stability of the abaltion front. Appl Phys A 57: 449-455

143. Luk'yanchuk B, Bityurin N, Anisimov S, Arnold N, Baeuerle D (1996a) The role of excited species in ultraviolet-laser materials ablation III. Non-stationary ablation of organic polymers. Appl. Phys. A 62: 397-401

144. Luk'yanchuk B, Bityurin N, Anisimov S, Malyshev A, Arnold N, Baeuerle D (1996b) Photophysical ablation of organic polymers: the influence of stresses. Applied Surface Science 106: 120-125

145. Luk'yanchuk B, Bityurin N, Himmelbauer M, Arnold N (1997) UV-lasser ablation of polyimide: From long to ultra-short laser pulses. Nucl Instr Meth Phys Res B 122: 347-355

146. Luk'yanchuk B, Bityurin N, Malyshev A, Anisimov S, Arnold N, Baeuerle D (1998) Photophysical ablation. SPIE 3343: 58-68

147. Mahan GD, Cole HS, Liu YS, Philipp HR (1988) Theory of polymer ablation. Appl Phys Lett 53: 2377-2379

148. Majumdar A (1993) Microscale heat conduction in dielectric thin films. ASME Journal of Heat Transfer 115: 7-16

149. Marchuk GI (1989) Splitting and alternating direction methods. In: Ciarlet PG, Lions JL (eds) Handbook of numerical analysis, vol 1. North-Holland, New York, pp197-465

150. Marten F, Hamielec A (1979) High conversion diffusion-controlled polymerization. In: Henderson J, Bouton T (eds) Polymerization reactors and processes. ACS Symposium Series 104, pp 43-70

151. Maxwell J, (1996) Three-dimensional laser-induced pyrolytic modeling, growth rate control, and application to micro-scale prototyping, Ph.D. Thesis, Rensselaer Polytechnical Institute

152. McMordie RK (1962) Steady state conduction with variable thermal conductivity. J Heat Transfer Trans ASME 84: 92-93

153. Modest MF, Abakians H (1986) Evaporative cutting of a semi-infinite body with a moving CW laser. J Heat Transfer 108: 602-607

154. Modest MF (1996) Three-dimensional, transient model for laser machining of ablating/decomposing materials. Int J Heat Mass Transfer 39: 221-234

155. Modest MF (1997) Laser through-cutting and drilling models for ablating/decomposing materials. J Laser Applications 9: 137-145

156. Murata K (1985) Theoretical studies of the electron scattering effect on developed pattern profiles in x-ray lithography. J Appl Phys 57: 575-580.

157. Musil C, Melngailis J, Etchin S, Hayes S (1996) Dose rate effects in GaAs investigated by discrete pulsed implantation using focused ion beam. J Appl Phys 80: 3727-3733

158. Nagamuchi S, Ueda M, Sakakima H, Satomi M, Ishikawa J (1996) Giant magnetoresistance in Co/Cu multilayers fabricated by focused ion beam direct deposition. J Appl Phys 80: 4217-4219

159. Nakamoto T, Yamaguchi K, Abraha P (1996) Manufacturing of three-dimensional micro-parts by UV laser induced polymerization. J Micromech. Microeng 6: 240-253

160. Nassar R, Krishnan A (1998) Mathematical model for the optimization of laser photoablated microstructures. SPIE 3512: 181-188

161. Nassar R, Vasile M, Zhang W (1998) Mathematical modeling of focused ion beam microfabrication. J Vac Sci Technol B 16: 109-115

162. Nassar R, Vasile M, Maxwell J (1999) A mathematical model for optimizing a laser induced photopolymerization process. SPIE 3875: 124-132

163. Nassar R, Zhang C, Dai W, Lan. H, Maxwell J (2002) Mathematical modeling of three-dimensional laser chemical vapor deposition, Microelectronic Engineering 60: 395-408

164. Nassar R, Dai W, Chen Q (2002b) An axisymmetric numerical model for simulating kinetically-limited growth of a cylindrical rod in 3D laser-induced chemical vapor deposition. J Mater Sci Technol 18: 127-132

165. Nikawa K (1991) Application of focused ion beam techniques to failure analysis of very large scale integrations. J Vac Sci Technol B 9: 2566-2577

166. Nilson RH, Griffiths SK (2000) Acoustic agitation for enhanced development of LIGA PMMA resists. SPIE 4174: 66-76

167. Nilson RH, Griffiths SK, Ting A (2001) Modeling acoustic agitation for enhanced development of PMMA resists. Fourth International workshop on High-Aspect-Ratio Micro-Structure Technology. Forschungszentrum Karlsruhe GmbH, Karlsruhe, Germany: 157-158

168. Nilson RH, Griffiths SK (2002) Enhanced transport by acoustic streaming in deep trench-like cavities. Journal of the Electrochemical Society 149: G286-G296

169. Nobes MJ, Colligon JS, Carter G (1969) The equilibrium topography of sputtered amorphous solids. J Mater Sci 4: 730-733

170. Nyborg WL (1998) Acoustic Streaming, Chapter 7 of Nonlinear Acoustics, MF Hamilton and DT Blackstock, edt, Academic Press, London

171. O'Driscoll KF, Kuindersma ME (1994) Monte Carlo simulation of pulsed laser polymerization. Macromolecules. Theory and Simulation 3: 469-478
172. Olaj OF, Bitai I, Hinkelmann F (1987) Laser-flash-initiated polymerization as a tool of evaluating (individual) kinetic constants of free-radical polymerization 2. The direct determination of the rate constant of rate propagation. Makromolekulare Chemie 188: 1689-1702
173. Orloff J (1993) High resolution focused ion beams. Review of Scientific Instruments 64: 1105-1130
174. Osgood RM (1983) Laser microchemistry and its application to electron-device fabrication. Ann. Rev. Phys. Chem. 34: 77-101
175. Oster G, Yang NL (1968) Photopolymerization of vinyl monomers. Chem Rev 68: 125-151
176. Pettit GH, Sauerbrey R (1993) Pulsed ultraviolet laser ablation. Appl Phys A 56: 51-63
177. Prewett PD (1993) Focused ion beams-Microfabrication methods and applications. Vacuum 44: 345-352
178. Qiu TQ, Tien CL (1992) Short-pulse laser heating on metals. International Journal of Heat and Mass Transfer 35: 719-726
179. Qiu TQ, Tien CL (1993) Heat transfer mechanisms during short-pulse laser heating of metals. ASME Journal of Heat Transfer 115: 835-841
180. Qiu TQ, Tien CL (1994) Femtosecond laser heating of multilayered metals- I. Analysis. International Journal of Heat and Mass Transfer 37: 2789-2797
181. Qiu TQ, Juhasz T, Suarez C, Bron WE, Tien CL (1994) Femtosecond laser heating of multi-layer metals II. Experiments. Int J Heat Mass Transfer 37: 2799-2808
182. Ross DS (1988) Ion etching: An application of the mathematical theory of hyperbolic conservation laws. J Electrochem Soc 135: 1235-1240
183. Roy S, Modest MF (1993) CW laser machining of hard ceramics I. Effects of three-dimensional conduction, variable properties and various laser parameters. Int J Heat Mass Transfer 36: 3515-3528
184. Sauerbrey R, Pettit GH (1989) Theory for the etching of organic materials by ultraviolet laser pulses. Appl Phys Lett 55: 421-423
185. Schmidt H, Ihlermann J, Wolff-Rottke B, Luther K, Troe J (1998) Ultraviolet laser ablation of polymers: spot size, pulse duration, and plume attenuation effects explained. J Appl Phys 83: 5458-5468
186. Seidel R, Howell JR (1992) Thermal Radiation Heat Transfer, 3rd edn. Hemisphere Publishing Corp, Boston
187. Sethian JA, (1996) Level Set Methods. Cambridge University Press, Cambridge
188. Shareef I, Maldonado JR (1989) Thermal and mechanical model of x-ray lithography masks under short pulse irradiation. J Vac Sci Technol B 7: 1575-1579
189. Shareef A, Maldonado JR, Vladimirsky Y, Katcoff DL (1990) Thermoelastic behavior of x-ray lithography masks during irradiation. IBM J Res Develop 34: 718-735
190. Sigmund P (1969) Theory of sputtering. I. Sputtering yield of amorphous and polycrystalline targets. Physical Review 184: 383-416
191. Singleton DL, Paraskevopoulos G, Taylor RS (1990) Comparison of theoretical models of laser ablation of polyimide with experimental results. Chem Phys 144: 415-423

192. Skouby DC, Jensen KF (1988) Modeling of pyrolitic laser-assisted chemical vapor deposition: Mass transfer and kinetic effects influencing the shape of the deposit. J Appl Phys 63: 198-206

193. Smith AN, Hostetler JL, Norris PM (1999) Nonequilibrium heating in metal films: An analytical and numerical analysis. Numerical Heat Transfer A 35: 859-873

194. Smith R, Walls JM (1980) Development of a general three-dimensional surface under ion bombardment. Philosophical Magazine A 42: 235-248

195. Smith R, Tagg MA, Carter G, Nobes MJ (1986) Erosion of corners and edges on an ion-bombarded silicon surface. J Mater Sci Lett 5: 115-120

196. Sommerfeld A (1954) Optics. Academic Press Inc, New York

197. Srinivasan R, Dreyfus RW (1985) Laser-induced fluorescence studies on ultraviolet laser ablation of polymers. Springer Ser Opt Sci 49: 396-400

198. Srinivasan V, Smrtic MA, Babu SV (1986) Excimer laser etching of polymers. J Appl Phys 59: 3861-3867

199. Srinivasan R, Hall RR, Loehle WD, Wilson WD, Allbee DC (1995) Chemical transformations of the polyimide Kapton brought about by ultraviolet laser radiation. J Appl Phys 78: 4881-4887

200. Sutcliffe E, Srinivasan R (1986) Dynamics of UV laser ablation of organic polymer surfaces. J Appl Phys 60: 3315-3322

201. Takagi T, Nakajima N (1993) Photoforming applied to fine machining. In: IEEE PROC of MEMS'93, pp173-178

202. Takagi T, Nakajima N (1994) Architecture combination by micro photoforming process. In: IEEE PROC of MEMS'94, pp 211-216

203. Tang DW, Araki N (1999) Wavy, wavelike, diffusive thermal responses of finite rigid slabs to high-speed heating of laser-pulses. Int J Mass Heat Transfer 42: 855-860

204. Taylor RS, Singleton DL, Paraskevopoulus G (1987) Effect of optical pulse duration on the XeCl laser ablation of polymers and biological tissue. Appl Phys Lett 50: 1779-1781

205. Tien CL, Chen G (1994) Challanges in microscale conductive and radiative heat transfer. ASME Journal of Heat Transfer 116: 799-807

206. Tokarev VN, Lunney JG, Marine W, Sentis M (1995) Analytical thermal model of ultraviolet laser ablation with single-photon absorption in the plume. J Appl Phys 78: 1241-1246

207. Treyz GV, Scarmozzino R, Osgood RM (1989) Deep ultraviolet laser etching of vias in polyimide films. Appl Phys Lett 55: 346-348

208. Tryson GR, Shultz AR (1979) A calorimetric study of acrylate photopolymerization. Journal of Polymer Science 17: 2059-2075

209. Tzou DY (1995a) A unified field approach for heat conduction from micro-to macro-scale. ASME Journal of Heat Transfer 117: 8-16

210. Tzou DY (1995b) The generalized lagging response in small-scale and high-rate heating. International Journal of Heat and Mass Transfer 38: 3231-3240

211. Tzou DY (1995c) Experimental support for the lagging responses in heat propagation. AIAA Journal of Thermophysics and Heat Transfer 9: 686-693

212. Tzou DY (1997) Macro to microscale heat transfer: The lagging behavior. Taylor and Francis, Washington DC

213. Vasile M, Harriott L (1989) Focused ion beam simulated deposition of organic compounds. J Vac Sci Technol B 7: 1954-1958

214. Vasile M, Grigg D, Griffith D, Fitzgerald J, Russell E (1991) Scanning probe tips formed by focused ion beams. Rev Sci Instrum 62: 2167-2171

215. Vasile M, Nassar R, Xie J (1998) Focused ion beam technology applied to microstructure fabrication. J Vac Sci Technol B 16: 2499-2505

216. Vasile M, Nassar R, Xie J, Guo H (1999a) Microfabrication techniques using focused ion beams and emergent applications. Micron 30: 235-244

217. Vasile M, Xie J, Nassar R (1999b) Depth control of focused-ion beam milling from a numerical model of the sputter process. J Vac Sci Technol B 17: 3085-3090

218. Vernotte P (1958) Les Paradoxes de la theorie continue de l'equation De la chaleur. Compte Rendus 246: 3154-3155

219. Vladimirsky Y, Maldonado J, Fair R, Acosta R, Vladimirsky O, Viswanathan R, Voelker H, Cerrina F, Wells GM, Hansen, Nachman MR (1989) Thermal effects in x-ray masks during synchrotron storage ring irradiation. J Vac Sci Technol B 7: 1657-1661

220. Vrentas JS, Duda JL (1977a) Diffusion in polymer em dash solvent systems em dash 1. Reexamination of the free-volume theory. J Polymer Sci 15: 403-416

221. Vrentas JS, Duda JL (1977b) Diffusion in polymer-solvent systems em dash 2. A predictive theory for the dependence of diffusion coefficients on temperature, concentration, and molecular weight. J Polymer Sci15: 417-439

222. Wagner RE (1974) Laser drilling mechanics. J Applied Physics 45: 4631-4637

223. Weast RC (1985) CRC handbook of chemistry and physics, 66th edn. CRC Press, Boca Raton, Fla., pp B9, B34, D178-179, E11

224. Whitham GB (1974) Linear and non-linear waves. Wiley, New York

225. Wise KD, Najafi K (1991) Microfabrication techniques for integrated sensors and microsystems. Science 254: 1335-1342

226. Xie J (1999) FIB microstructure fabrication with mathematical modeling. Ph.D. Dissertation, College of Engineering and Science, Louisiana Tech University

227. Ximen H, Defreez RK, Orloff J, Elliott RA, Evans GA, Carlson NW, Lurie M, Bour DP (1990) Focused ion beam micromachined three-dimensional features by means of a digital scan. J Vac Sci Technol B 8: 1361-1365

228. Yamamura Y, Itikawa Y, Itoh N (1983) Angular dependence of sputtering yields of monoatomic solids. In: IPPJ-AM-26. Institute of Plasma Physics, Nagoya University, Nagoya, Japan

229. Yamaguchi K, Nakamoto T, Abbay PA, Mibu S (1994a) The accuracy of micromolds produced by a UV laser induced polymerization. Transactions of the ASME 116: 370-376

230. Yamaguchi K, Nakamoto T, Petros A (1994b) Consideration on the accuracy of fabricating microstructures using UV laser induced polymerization. Proceedings of the 5th international symposium on micro-machine and human science. Nagoya, Japan, 171-176

231. Yamaguchi K, Nakamoto T, Abraha P (1995) Consideration on the optimum conditions to produce micromechanical parts by photo polymerization using direct focused beam writing. In: IEEE sixth international symposium on micro machine and human science. 71-76

232. Yankova MI, Copley SM (1993) Numerical method for modeling pyrolytic laser-assisted chemical vapor deposition. International Conference on Beam Processing of Advanced Materials, Singh J, Copley SM (eds),185-192

233. Young R (1993) Micro-machining using a focused ion beam.Vacuum 44: 353-356
234. Zel'dovich Ya B, Raizer Yu P (1967) Physics of Shock Waves, High-Temperature Hydrodynamic Phenomena. Academic Press, New York
235. Zumaque H, Kohring GA, Hormes J (1997) Simulation studies of energy deposition and secondary processes in deep x-ray lithography. J. Micromech. Microeng. 7: 79-88.

Index

Druck: Strauss Offsetdruck, Mörlenbach
Verarbeitung: Schäffer, Grünstadt